Board Review Series

Physiology

Look for other titles in this series:

Behavioral Science, *2nd edition*
Biochemistry, *2nd edition*
Cell Biology and Histology, *2nd edition*
Embryology
Gross Anatomy, *2nd edition*
Microbiology and Immunology, *2nd edition*
Neuroanatomy
Pathology
Pharmacology, *2nd edition*

Look for these forthcoming titles:

Gross Anatomy, *3rd edition*
Neuroanatomy, *2nd edition*

Board Review Series

Physiology

Linda S. Costanzo, Ph.D.
Professor of Physiology
Medical College of Virginia
Virginia Commonwealth University
Richmond, Virginia

Williams & Wilkins

Philadelphia • Baltimore • Hong Kong • London • Munich • Sydney • Tokyo

A Waverly Company

Williams & Wilkins

Acquisitions Editor: Elizabeth A. Nieginski
Managing Editor: Susan E. Kelly
Project Editor: Amy G. Dinkel
Illustration: Matthew C. Chansky
Production Manager: Laurie Forsyth

ISBN 0-683-2134-6

Printed in the United States of America

96 97 98
3 4 5 6 7 8 9 10

Dedication

For Richard
and
for Dan and Rebecca

Contents

Preface

The subject matter of physiology is the foundation of the practice of medicine, and a firm grasp of its principles is essential for the practicing physician. This book is intended to aid the student preparing for the United States Medical Licensing Examination (USMLE) Step 1. It is a concise review of key physiologic principles and is intended to help the student recall material taught during the first and second years of medical school. It is not intended to substitute for comprehensive textbooks or for course syllabi, although the student may find it a useful adjunct to physiology and pathophysiology courses in the first and second years of medical school.

Organization of the book

The material is organized by organ system into seven chapters. The first chapter reviews general principles of cellular physiology. The remaining six chapters review the major organ systems—neurophysiology, cardiovascular, respiratory, renal and acid–base, gastrointestinal, and endocrine physiology. Because the study of physiology builds upon itself, most students will wish to use the book in the order of presentation.

Difficult concepts are explained step-wise, concisely and clearly, with appropriate illustrative examples and sample problems. Numerous clinical correlations are included so that the student can understand physiology and its relationship to medicine. An integrative approach is used to demonstrate physiologic concepts and how the organ systems work together to maintain homeostasis. There are 114 illustrations and flow charts and 46 tables to help the student visualize the material quickly and to aid in long-term retention.

Questions reflecting the content and format of USMLE Step 1 are included at the end of each chapter and in a Comprehensive Examination at the end of the book. These questions, many with clinical relevance, require problem-solving skills. Clear, concise explanations accompany the questions and guide the student through the correct steps of reasoning. The questions can be used as a pretest to identify areas of weakness or as a posttest to determine mastery of the subject matter; when using the questions as a posttest, the student should *not* do so immediately after studying the chapter. Special attention should be given to the Comprehensive Examination—these questions integrate several areas of physiology as well as related concepts of pathophysiology and pharmacology.

I wish you the best of luck in your preparation for USMLE Step 1.

Linda S. Costanzo, Ph.D.

Acknowledgments

My sincere thanks to the entire staff at Williams & Wilkins. Charles Duvall introduced me to Elizabeth Nieginski, who convinced me that I could write a physiology review book. Susan Kelly, Managing Editor of the Board Review Series, supported me through each step in the process of studying, writing, reviewing, and editing; her extraordinarily critical eye and attentiveness to detail are evident throughout. Matthew Chansky created all of the illustrations, which complement the text; he made my life easy by understanding the need for simplicity and clarity. A review panel of medical school faculty and students offered many helpful suggestions that have been incorporated into the book. Laurie Forsyth, Production Manager, and Amy Dinkel, Project Editor, were a pleasure to work with.

Members of the faculty at the Medical College of Virginia shared their expertise with me; Drs. Clive Baumgarten, Peter Boling, Robert Downs, Douglas Heuman, James Levenson, Henry Rozycki, Paul Swerdlow, and Raphael Witorsch are explemplary colleagues who answered my questions and directed me to resources, clarified difficult concepts, and shared my enthusiasm for physiology. My husband, Dr. Richard Costanzo, tutored and coached me through the writing of the neurophysiology chapter and cheered me on through the entire process.

Students in the classes of 1994 and 1995 at the Medical College of Virginia critically reviewed sections of the book—Audrey Hernandez, Amir Jamali, Samir Mardini, Curtis Thwing, and Tom Thompson; Eugene Chang graciously reviewed the entire manuscript and met every deadline set for him. Finally, I offer thanks to all 170 members of the Class of 1996 who used a draft of this book to prepare for USMLE Step 1; they spotted errors at an early stage, gave me essential feedback on style and presentation, and, most of all, kept me on task by asking, "How's the book coming, Dr. C?"

Linda S. Costanzo

1
Cell Physiology

I. Cell Membranes

—are composed primarily of phospholipids and proteins.

A. Lipid bilayer

1. **Phospholipids** have a **glycerol backbone,** which is the hydrophilic (water-soluble) head, and two **fatty acid tails,** which are hydrophobic (water-insoluble). The hydrophobic tails face each other and form a bilayer.

2. **Lipid-soluble substances** cross cell membranes because they can dissolve in the hydrophobic lipid bilayer. **Water-soluble substances** cannot dissolve in the lipid of the membrane, but must cross through water-filled channels or pores or may be transported by carriers.

B. Proteins

1. **Integral proteins**
   —span the entire membrane.
   —are anchored through hydrophobic interactions with the phospholipid bilayer.
   —**For example:** an **ion channel**.

2. **Peripheral proteins**
   —are located on either the intracellular or the extracellular side of the cell membrane.

C. Intercellular connections

1. **Tight junctions (zonula occludens)**
   —are the attachments between cells (often epithelial cells).
   —may be an intercellular pathway for solutes, depending on the size, charge, and characteristics of the tight junction.
   —may be **"tight"** (impermeable) as in the renal distal tubule, or **"leaky"** (permeable) as in the renal proximal tubule or gallbladder.

1

2. Gap junctions
  —are the attachments between cells that permit intercellular communication.
  —**For example,** gap junctions permit current flow and electrical coupling between myocardial cells.

II. Transport Across Cell Membranes (Table 1-1)

A. Simple diffusion

1. Characteristics of simple diffusion
  —is the only form of transport across cell membranes that is **not carrier-mediated**.
  —occurs **down an electrochemical gradient** ("downhill").
  —does not require metabolic energy and therefore is **passive**.

2. Diffusion can be measured using the following equation:

$$\mathbf{J} = -\mathbf{P\,A\,(C_1 - C_2)}$$

J	flux (flow)	mmol/sec
A	area	cm^2
P	permeability	cm/sec
C$_1$	concentration$_1$	mmol/L
C$_2$	concentration$_2$	mmol/L

3. Sample calculation for diffusion
  —The urea concentration of blood is 10 mg/100 ml and the urea concentration of proximal tubular fluid is 20 mg/100 ml. If the permeability to urea is 1×10^{-5} cm/sec and surface area is 100 cm^2, what is the magnitude and direction of the urea flux?

Table 1-1. Characteristics of Different Types of Transport

Type	Electrochemical Gradient	Carrier-mediated	Metabolic Energy	Na$^+$ Gradient	Inhibition of Na$^+$–K$^+$ Pump
Simple diffusion	Downhill	No	No	No	
Facilitated diffusion	Downhill	Yes	No	No	
Primary active transport	Uphill	Yes	Yes		Inhibits (if Na$^+$–K$^+$ pump)
Cotransport	Uphill*	Yes	Indirect	Yes, same direction	Inhibits
Countertransport	Uphill*	Yes	Indirect	Yes, opposite direction	Inhibits

*One or more solutes are transported uphill; Na$^+$ is transported downhill.

−**Note:** The minus sign indicates that the direction of flux is from high to low concentration; it can be ignored if the higher concentration is called C_1 and the lower concentration is called C_2. **Also note:** 1 ml = 1 cm^3.

$$\text{Flux} = \left(\frac{1 \times 10^{-5} \text{ cm}}{\text{sec}} \right) (100 \text{ cm}^2) \left(\frac{20 \text{ mg}}{100 \text{ ml}} - \frac{10 \text{ mg}}{100 \text{ ml}} \right)$$

$$= \left(\frac{1 \times 10^{-5} \text{ cm}}{\text{sec}} \right) (100 \text{ cm}^2) \left(\frac{10 \text{ mg}}{100 \text{ ml}} \right)$$

$$= \left(\frac{1 \times 10^{-5} \text{ cm}}{\text{sec}} \right) (100 \text{ cm}^2) \left(\frac{0.1 \text{ mg}}{\text{cm}^3} \right)$$

$$= 1 \times 10^{-4} \text{ mg/sec from lumen to blood (high to low concentration)}$$

4. Permeability

−is the P in the equation for diffusion.
−describes the ease with which a solute diffuses through a membrane.
−depends on the characteristics of the solute and the membrane.
−**Factors that increase permeability:**

a. ↑ **oil/water partition coefficient of the solute** increases solubility in the lipid of the membrane.

b. ↓ **radius (size) of the solute** increases the speed of diffusion.

c. ↓ **membrane thickness** decreases the diffusion distance.

−Small hydrophobic solutes have the highest permeabilities in lipid membranes.

−Hydrophilic solutes must cross cell membranes through water-filled channels or pores. **If the solute is an ion (is charged), then its flux will depend on both the concentration difference and the potential difference across the membrane.**

B. Characteristics of carrier-mediated transport

−apply to facilitated diffusion and primary and secondary active transport:

1. **Stereospecificity. For example,** D-glucose (the natural isomer) is transported by facilitated diffusion, but the L-isomer is not. Simple diffusion, however, would not distinguish between the two isomers because it does not involve a carrier.

2. **Saturation.** Transport rate increases as the concentration of the solute increases, until the carriers are saturated. The transport maximum **(T_m)** **is analogous to the V_{max} in enzyme kinetics.**

3. **Competition.** Structurally related solutes compete for transport sites on carrier molecules. **For example,** galactose is a competitive inhibitor of glucose transport in the small intestine.

C. Facilitated diffusion

1. Characteristics of facilitated diffusion

−occurs **down an electrochemical gradient** ("downhill"), like simple diffusion.
−does not require metabolic energy and therefore is **passive.**
−is more **rapid** than simple diffusion.

–is **carrier-mediated** and therefore exhibits stereospecificity, saturation, and competition.

2. Example of facilitated diffusion

–Glucose transport in muscle and adipose cells is "downhill," is carrier-mediated, and is inhibited by sugars such as galactose; therefore, it qualifies as facilitated diffusion. In **diabetes mellitus,** glucose uptake and utilization by muscle and adipose cells is impaired because the carriers for facilitated diffusion of glucose require **insulin.**

D. Primary active transport

1. Characteristics of primary active transport

–occurs **against an electrochemical gradient** ("uphill").

–requires **direct input of metabolic energy** in the form of adenosine triphosphate (ATP) and so is **active.**

–is **carrier-mediated** and so exhibits stereospecificity, saturation, and competition.

2. Examples of primary active transport

a. Na^+–K^+ ATPase (or Na^+–K^+ pump) in cell membranes transports Na^+ from intracellular to extracellular fluid and K^+ from extracellular to intracellular fluid; it maintains low intracellular $[Na^+]$ and high intracellular $[K^+]$.

–Both **Na^+ and K^+ are transported against their electrochemical gradients.**

–Energy is provided from the terminal phosphate bond of ATP.

–The **usual stoichiometry** is 3 Na^+/2 K^+.

–Specific inhibitors of the Na^+–K^+ ATPase are **ouabain** and **digitalis.**

b. Ca^{2+}–ATPase (or Ca^{2+} pump) in sarcoplasmic reticulum or cell membranes transports Ca^{2+} against an electrochemical gradient.

c. K^+–H^+–ATPase (or proton pump) in stomach parietal cells transports H^+ into the lumen of the stomach against its electrochemical gradient.

–It is inhibited by **omeprazole.**

E. Secondary active transport

1. Characteristics of secondary active transport

a. The transport of two or more solutes is **coupled.**

b. One of the solutes (usually Na^+) is transported "downhill" and provides energy for the "uphill" transport of the other solute(s).

c. Metabolic energy is not provided directly, but indirectly from the Na^+ gradient, which is maintained across cell membranes. Thus, inhibition of the Na^+–K^+ ATPase will decrease transport of Na^+ out of the cell, decrease the transmembrane Na^+ gradient, and eventually inhibit secondary active transport.

d. If the solutes move in the same direction across the cell membrane, it is called **cotransport** or **symport.**

–**Examples are** Na^+–glucose cotransport in the small intestine and Na^+–K^+–Cl^- cotransport in the renal thick ascending limb.

    e. If the solutes move in opposite directions across the cell membranes, it is called **countertransport** or **exchange** or **antiport**.

–**Examples** are Ca^{2+}–Na^+ exchange and Na^+–H^+ exchange.

2. Example of Na^+–glucose cotransport (Figure 1-1)

    a. The carrier for Na^+–glucose cotransport is located in the luminal membrane of intestinal mucosal and renal proximal tubule cells.

    b. Glucose is transported uphill, and Na^+ is transported downhill.

    c. Energy is derived from the downhill movement of Na^+. The Na^+ gradient is maintained by the Na^+–K^+ pump on the basolateral (blood side) membrane. Poisoning the Na^+–K^+ pump decreases the transmembrane Na^+ gradient and consequently inhibits Na^+–glucose cotransport.

3. Example of Ca^{2+}–Na^+ countertransport or exchange (Figure 1-2)

    a. Many cell membranes contain a Ca^{2+}–Na^+ exchanger that transports Ca^{2+} "uphill" from low intracellular $[Ca^{2+}]$ to high extracellular $[Ca^{2+}]$. Ca^{2+} and Na^+ move in opposite directions across the cell membrane.

    b. The energy is derived from the "downhill" coupled transport of Na^+. As with cotransport, the inwardly directed Na^+ gradient is maintained by the Na^+–K^+ pump. Poisoning the pump inhibits Ca^{2+}–Na^+ exchange.

III. Osmosis

A. Osmolarity

–is the concentration of osmotically active particles in a solution.
–is a colligative property that can be measured by freezing point depression.
–can be calculated using the following **equation:**

Osmolarity $=$ **g** $\times$ **C**

osmolarity	**concentration of particles**	**osmol/L**
C	**concentration**	**mol/L**
g	**number of particles in solution**	**osmol/mol**

Ex: $g_{NaCl} = 2$
$g_{glucose} = 1$

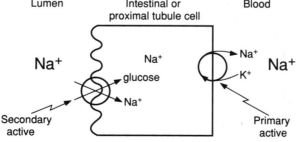

Figure 1-1. Na^+–glucose cotransport (symport) in intestinal or proximal tubule epithelial cell.

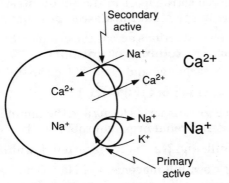

Figure 1-2. Ca^{2+}–Na^+ countertransport (antiport).

—Two solutions having the same calculated osmolarity are **isosmotic**. If two solutions have different calculated osmolarities, the solution with the higher osmolarity is **hyperosmotic** and the solution with the lower osmolarity is **hyposmotic**.

—**Sample calculation:** What is the osmolarity of a 1 M NaCl solution?

$$\text{Osmolarity} = g \times C$$
$$= 2 \text{ osmol/mol} \times 1 \text{ M}$$
$$= 2 \text{ osmol/L}$$

B. Osmosis and osmotic pressure

—**Osmosis is the flow of water** across a semipermeable membrane from a solution with low solute concentration to a solution with high solute concentration.

1. **Example of osmosis** (Figure 1-3)

   a. Solutions 1 and 2 are separated by a semipermeable membrane. Solution 1 contains a solute that is too large to cross the membrane. Solution 2 is pure water. The presence of the solute in solution 1 produces an **osmotic pressure**.

   b. The osmotic pressure difference across the membrane causes water to flow from solution 2 (which has no solute and the lower osmotic pressure) to solution 1 (which has the solute and the higher osmotic pressure).

   c. With time, the volume of solution 1 increases and the volume of solution 2 decreases.

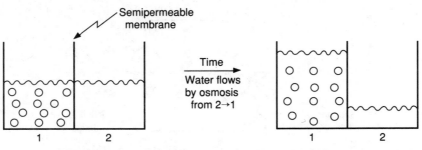

Figure 1-3. Osmosis of H_2O across a semipermeable membrane.

2. Calculating osmotic pressure (van't Hoff's law)

a. The **osmotic pressure** of solution 1 (see Figure 1-3) can be calculated by van't Hoff's law, which states that osmotic pressure depends on the concentration of osmotically active particles. Concentration of particles is converted to pressure according to the following **equation**:

$$\pi = g \times RT \times C$$

π	osmotic pressure	mm Hg or atm
g	number of particles in solution	osmol/mol
R	gas constant	0.082 L – atm/mol – °K
T	absolute temperature	°K
C	concentration	mol/L

b. **The osmotic pressure increases when the solute concentration increases.** A solution of 1 M $CaCl_2$ has a higher osmotic pressure than a solution of 1 M KCl because the concentration of particles is higher.

c. The higher the osmotic pressure of a solution, the greater the water flow into it.

d. Two solutions having the same effective osmotic pressure are **isotonic** since no water will flow across a semipermeable membrane separating them. If two solutions separated by a semipermeable membrane have different effective osmotic pressures, the solution with the higher osmotic pressure is **hypertonic** and the solution with the lower osmotic pressure is **hypotonic**. Water flows from the hypotonic to the hypertonic solution.

e. **Colloid osmotic pressure** or **oncotic pressure** is the osmotic pressure created by protein (e.g., in blood).

3. Reflection coefficient

–is a number between zero and one that describes the ease with which a solute permeates a membrane.

–**If the reflection coefficient is one,** then the solute is impermeable, is retained in the original solution, creates an osmotic pressure, and causes water flow.

–**If the reflection coefficient is zero,** then the solute is completely permeable and, therefore, will not exert any osmotic effect, and will not cause water flow.

–**Serum albumin** (a large solute) has a reflection coefficient close to one; **urea** (a small solute) has a reflection coefficient close to zero.

IV. Diffusion Potential, Resting Membrane Potential, and Action Potential

A. Ion channels

–are **integral proteins** that span the membrane and permit, when open, the passage of certain ions.

1. Ion channels

–**are selective** and permit the passage of some ions but not others. Selectivity is based on the size of the channel and the distribution of charges lining it.

–**For example,** a small channel lined with negatively charged groups will be selective for small cations and exclude large solutes or anions.

2. Ion channels

–**may be open or closed.** When the channel is open, the ion(s) for which it is selective can flow through. When it is closed, ions cannot flow through.

3. The **conductance of a channel** depends upon the probability that the channel is open. The higher the probability that a channel is open, the higher the conductance or **permeability**. Opening and closing of channels is controlled by **gates**.

    a. Voltage-gated channels are opened or closed by changes in membrane potential.

       –**For example,** the activation gate of the Na^+ channel in nerve is opened by depolarization; when open, the nerve membrane is permeable to Na^+ (compare the upstroke of the nerve action potential).

    b. Ligand (chemically)-gated channels are opened or closed by chemicals such as hormones, second messengers, and neurotransmitters.

       –**For example,** the nicotinic receptor for acetylcholine (ACh) at the motor end plate is an ion channel that opens when ACh binds to it. When open, it is permeable to Na^+, K^+, and Ca^{2+}, causing the motor end plate to depolarize.

B. Diffusion and equilibrium potentials

–A **diffusion potential** is the potential difference generated across a membrane because of a concentration difference of an ion. A diffusion potential can be generated only if the membrane is permeable to the ion.

–The **size of the diffusion potential** depends on the size of the concentration gradient.

–The **sign of the diffusion potential** depends on whether the diffusing ion is positively or negatively charged.

–Diffusion potentials are created by the diffusion of **very few ions** and so concentrations in the bulk solutions do not change.

–The **equilibrium potential** is the diffusion potential that exactly balances (opposes) the tendency for diffusion caused by a concentration difference. At **electrochemical equilibrium,** the chemical and electrical driving forces acting on an ion are equal and opposite and there is no more net movement.

1. Example of a Na^+–diffusion potential (Figure 1-4)

    a. Suppose that two solutions of NaCl are separated by a membrane that is permeable to Na^+ but not to Cl^-. The NaCl concentration of solution 1 is higher than that of solution 2.

    b. Because the membrane is permeable to Na^+, it will diffuse from solution 1 to solution 2 down its concentration gradient. Cl^- will not accompany Na^+.

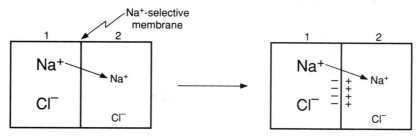

Figure 1-4. Generation of Na^+-diffusion potential across Na^+-selective membrane.

    c. As a result, a **diffusion potential** will develop such that solution 1 becomes negative with respect to solution 2.

    d. Eventually, the potential difference will become large enough to oppose further net diffusion of Na^+. The potential difference that exactly balances the diffusion of Na^+ is the Na^+ **equilibrium potential**. At electrochemical equilibrium the chemical and electrical driving forces on Na^+ are equal and opposite.

2. Example of a Cl^-–diffusion potential (Figure 1-5)

    a. Suppose that solutions identical to those in Figure 1-4 are now separated by a membrane that is permeable to Cl^- rather than to Na^+.

    b. Cl^- will diffuse from solution 1 to solution 2 down its concentration gradient. Na^+ will not accompany Cl^-.

    c. A diffusion potential will be established so that solution 1 becomes positive with respect to solution 2. The potential difference that exactly balances the diffusion of Cl^- down its concentration gradient is the Cl^- **equilibrium potential**. At electrochemical equilibrium, the chemical and electrical driving forces on Cl^- are equal and opposite.

3. Using the Nernst equation to calculate equilibrium potentials

    a. The **Nernst equation** is used to calculate the equilibrium potential for a concentration difference of a permeant ion across a cell membrane. It tells us what potential would exactly balance the tendency for diffusion down the concentration gradient; in other words, **at what potential would the ion be at electrochemical equilibrium?**

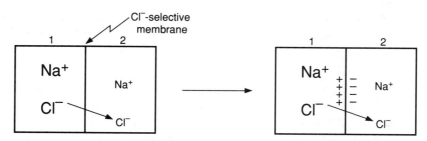

Figure 1-5. Generation of a Cl^--diffusion potential across a Cl^--selective membrane.

$$E = \frac{-2.3\,RT}{zF}\ \log_{10}\ \frac{[C_i]}{[C_e]}$$

E	equilibrium potential	mV
$\dfrac{2.3\,RT}{F}$	constants	60 mV
z	charge on the ion	$+1$ for Na^+
		$+2$ for Ca^{2+}
		-1 for Cl^-
C_i	intracellular concentration	mM
C_e	extracellular concentration	mM

b. Sample calculation with the Nernst equation

—If the intracellular $[Na^+]$ is 15 mM and the extracellular $[Na^+]$ is 150 mM, what is the equilibrium potential for Na^+?

$$\begin{aligned} E_{Na^+} &= \frac{-60\ mV}{z}\ \log_{10}\ \frac{[C_i]}{[C_e]} \\[2mm] &= \frac{-60\ mV}{+1}\ \log_{10}\ \frac{15\ mM}{150\ mM} \\[2mm] &= -60\ mV\quad \log_{10}\ 0.1 \\[2mm] &= +60\ mV \end{aligned}$$

Note: You need not remember which concentration goes in the numerator. Because it is a log function, do the calculation either way to get 60 mV, and then get the correct sign with an "intuitive approach." (Intuitive approach: The $[Na^+]$ is higher in extracellular fluid than in intracellular fluid, so Na^+ ions will tend to diffuse from extracellular to intracellular, making the inside of the cell positive [i.e., $+60$ mV at equilibrium].)

c. Typical values for equilibrium potentials in nerve and muscle

E_{Na^+}	$+65$ mV
$E_{Ca^{2+}}$	$+120$ mV
E_{K^+}	-85 mV
E_{Cl^-}	-90 mV

C. Resting membrane potential

—is the measured potential difference across the cell membrane in millivolts (mV).

—is the intracellular potential relative to the extracellular potential. Thus, a resting membrane potential of -70 mV means **70 mV, cell negative**.

1. It is established by diffusion potentials resulting from concentration differences of permeant ions.

2. Each **permeant ion** attempts to drive the membrane potential toward **its equilibrium potential**. Ions with the highest permeabilities or conductances will make the greatest contributions to the resting membrane potential, and those with the lowest permeabilities will make little or no contribution.

3. The resting membrane potential of nerve is -70 mV, close to the calculated K^+ equilibrium potential of -85 mV, but far from the calculated Na^+ equilibrium potential of $+65$ mV. **At rest, the nerve membrane is far more permeable to K^+ than to Na^+.**

4. The Na^+–K^+ pump contributes indirectly to the resting membrane potential by maintaining, across the cell membrane, the Na^+ and K^+ concentration gradients that then produce diffusion potentials. The direct **electrogenic** contribution of the pump (because it pumps 3 Na^+/2 K^+) is small.

D. Action potentials

1. Definitions

  a. **Depolarization** makes the membrane potential less negative. **Hyperpolarization** makes the membrane potential more negative.

  b. **Inward current** is the flow of positive charge into the cell. Inward currents **depolarize** the membrane potential.

  c. **Outward current** is the flow of positive charge out of the cell. Outward currents **hyperpolarize** the membrane potential.

  d. **Action potential** is a property of excitable cells (nerve, muscle) consisting of a rapid depolarization followed by repolarization of the membrane potential. Action potentials have **stereotypical size and shape,** are **propagating,** and are **all-or-none.**

  e. **Threshold** is the membrane potential at which occurrence of the action potential is inevitable. Inward currents depolarize the membrane to threshold. Subthreshold inward currents do not bring the membrane to threshold and do not produce an action potential.

2. Ionic basis of the nerve action potential (Figure 1-6)

  a. **Resting membrane potential**
   –is approximately -70 mV, cell negative.
   –is the result of the **high resting conductance to K^+,** which drives the membrane potential toward the K^+ equilibrium potential.
   –At rest, the Na^+ channels are closed and Na^+ conductance is low.

  b. **Upstroke of action potential**
   (1) Inward current depolarizes the membrane potential to threshold.
   (2) **Depolarization causes rapid opening of the activation gates of the Na^+ channel,** and the Na^+ conductance of the membrane promptly increases.
   (3) The Na^+ conductance becomes higher than the K^+ conductance and so the membrane potential is driven toward (but does not quite reach) the Na^+ equilibrium potential of $+65$ mV. **Note:** The rapid depolarization during the upstroke is caused by an **inward Na^+ current.**
   (4) The **overshoot** is the portion of the action potential when the membrane potential is positive.
   (5) **Tetrodotoxin (TTX)** blocks these voltage-sensitive Na^+ channels and so abolishes action potentials.

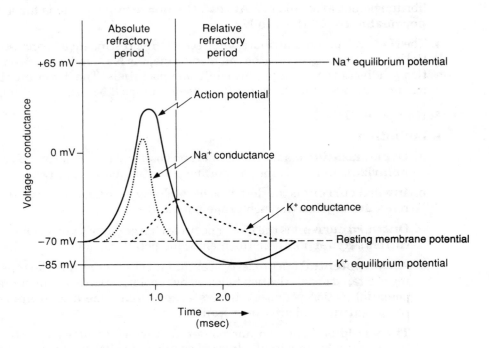

Figure 1-6. Nerve action potential and associated changes in Na$^+$ and K$^+$ conductance.

c. Repolarization of action potential

    (1) Depolarization also closes the inactivation gates of the Na$^+$ channel (but more slowly than it opens the activation gates). Closure of the inactivation gates means that the Na$^+$ channels close, and so Na$^+$ conductance returns toward zero.

    (2) Depolarization slowly opens K$^+$ channels and increases K$^+$ conductance to even higher levels than at rest.

    (3) The **combined effect** of closing the Na$^+$ channels and greater opening of the K$^+$ channels makes the K$^+$ conductance higher than the Na$^+$ conductance, and the membrane potential is repolarized. **Note:** The repolarization is caused by an **outward K$^+$ current**.

d. Undershoot (hyperpolarizing afterpotential)

    —The K$^+$ conductance remains high for some time after closure of the Na$^+$ channels. During this period, the membrane potential is driven very close to the K$^+$ equilibrium potential.

3. Refractory periods (see Figure 1-6)

a. Absolute refractory period

    —is the period during which another action potential cannot be elicited, no matter how large the stimulus.

    —coincides with almost the entire duration of the action potential.

—**Explanation:** The inactivation gates of the Na^+ channel are closed and will remain closed until repolarization occurs. No action potential can occur until the inactivation gates open.

   b. **Relative refractory period**

—begins at the end of the absolute refractory period and continues until the membrane potential returns to the resting level.

—An action potential can be elicited during this period if a stronger than usual current is provided.

—**Explanation:** The K^+ conductance is higher than at rest, the membrane potential is closer to the K^+ equilibrium potential and farther from threshold; more current is required to bring the membrane to threshold.

  4. Propagation of action potentials

—occurs by spread of **local currents** to adjacent areas of membrane, which are then depolarized to threshold and generate action potentials.

—**Conduction velocity is increased by:**

   a. ↑ **fiber size.** Increasing diameter of a nerve fiber results in decreased internal resistance and so conduction velocity down the nerve is faster.

   b. **Myelination.** Myelin acts as an insulator around nerve axons and increases conduction velocity. Myelinated nerves exhibit **saltatory conduction** because action potentials can be generated only at the **nodes of Ranvier,** where there are gaps in the myelin sheath.

V. Neuromuscular and Synaptic Transmission

A. General characteristics of chemical synapses

1. An action potential in the presynaptic cell causes depolarization of the presynaptic terminal.

2. As a result of the depolarization, Ca^{2+} **enters the presynaptic terminal.** Ca^{2+} entry causes **release of neurotransmitter into the synaptic cleft.**

3. Neurotransmitter diffuses across the synaptic cleft and combines with **receptors on the postsynaptic cell membrane,** causing a change in its permeability to ions and its membrane potential.

4. Inhibitory neurotransmitters hyperpolarize the postsynaptic membrane; **excitatory neurotransmitters** depolarize the postsynaptic membrane.

B. Neuromuscular junction (Table 1-2)

—is the synapse between axons of motoneurons and skeletal muscle.

—The neurotransmitter released from the presynaptic terminal is **ACh,** and the receptor on the postsynaptic membrane is **nicotinic.**

1. Synthesis and storage of ACh in the presynaptic terminal

—**Choline acetyltransferase** catalyzes the formation of ACh from acetyl coenzyme A (CoA) and choline in the presynaptic terminal.

—ACh is stored in **synaptic vesicles** with ATP and proteoglycan for later release.

Table 1-2. Agents Affecting Neuromuscular Transmission

Example	Action	Effect on Neuromuscular Transmission
Botulinus toxin	Blocks release of ACh from presynaptic terminals	Total blockade
Curare	Competes with ACh for receptors on motor end plate	Decreases size of EPP; maximal doses produce paralysis of respiratory muscles and death
Neostigmine	Anticholinesterase	Prolongs and enhances action of ACh at muscle end plate
Hemicholinium	Blocks reuptake of choline into presynaptic terminal	Depletes ACh stores from presynaptic terminal

2. Depolarization of the presynaptic terminal and Ca^{2+} uptake

—Action potentials are conducted down the motoneuron to its presynaptic terminal where **depolarization opens Ca^{2+} channels**.

—When the Ca^{2+} permeability increases, Ca^{2+} rushes into the presynaptic terminal down its electrochemical gradient.

3. Ca^{2+} uptake causes release of ACh into the synaptic cleft

—The synaptic vesicles fuse with the plasma membrane and empty their contents into the cleft by **exocytosis**.

4. Diffusion of ACh to the postsynaptic membrane (muscle end plate) and binding to specific receptors

—The ACh receptor is also a **Na^+ and K^+ ion channel**.

—Binding of ACh to its α subunits causes a conformational change that opens the central core of the channel and increases its conductance to Na^+ and K^+. These are **ligand-gated channels**.

5. End plate potential (EPP) in the postsynaptic membrane

—Because the channels opened by ACh conduct both Na^+ and K^+ ions, the postsynaptic membrane potential is depolarized to a value halfway between their respective equilibrium potentials, about 0 mV.

—The contents of one synaptic vesicle (one quantum) produces a **miniature end plate potential** (MEPP—the smallest possible EPP).

—MEPPs summate to produce a full-fledged EPP. **The EPP is not an action potential,** but a depolarization of the specialized muscle end plate.

6. Depolarization of adjacent muscle membrane to threshold

—Once the end plate region is depolarized, local currents cause depolarization and action potentials in the adjacent muscle. Contraction follows.

7. Degradation of ACh

—The EPP is transient because ACh is degraded to acetyl CoA and choline by **acetylcholinesterase** (AChE) on the muscle end plate.

—One-half of the choline is taken back into the presynaptic ending by Na^+-choline cotransport.

—**AChE inhibitors (neostigmine)** block the degradation of ACh, prolong its action at the muscle end plate, and increase the size of the EPP.

—**Hemicholinium** blocks choline reuptake and so depletes the presynaptic endings of ACh stores.

8. Disease—myasthenia gravis

—is characterized by skeletal muscle weakness and fatigability resulting from a **reduced number of ACh receptors** on the muscle end plate (caused by circulating antibodies to the receptors).

—The size of the EPP is reduced and, therefore, it is harder to depolarize the muscle membrane to threshold to produce action potentials.

—**Treatment with AChE inhibitors** prolongs the action of ACh at the muscle end plate and partly compensates for the reduced number of receptors.

C. Synaptic transmission

1. Types of arrangements

a. One-to-one synapses (such as neuromuscular junction)

—An action potential in the presynaptic element (the motor nerve) produces an action potential in the postsynaptic element (the muscle).

b. Many-to-one synapses (such as spinal motoneurons)

—In these synapses, an action potential in a single presynaptic cell is insufficient to produce an action potential in the postsynaptic cell. Instead, many cells synapse on the postsynaptic cell to depolarize it to threshold. They may be excitatory or inhibitory.

2. Input to synapses

—The postsynaptic cell integrates excitatory and inhibitory inputs.

—When the sum of all the input brings its membrane potential to threshold, it fires an action potential.

a. Excitatory postsynaptic potentials (EPSP)

—are input that **depolarize** the postsynaptic cell, bringing it closer to threshold and closer to firing an action potential.

—are caused by **opening of channels that are permeable to Na^+ and K^+,** similar to the ACh channels. The membrane potential depolarizes to a value halfway between the equilibrium potentials for Na^+ and K^+, about 0 mV.

—**Excitatory neurotransmitters** include ACh, norepinephrine, epinephrine, dopamine, glutamate, and serotonin.

b. Inhibitory postsynaptic potentials (IPSP)

—are input that **hyperpolarize** the postsynaptic cell, moving it away from threshold and farther from firing an action potential.

—are caused by **opening Cl^- channels.** The membrane potential is hyperpolarized toward the Cl^- equilibrium potential (-90 mV).

—**Inhibitory neurotransmitters** are γ-aminobutyric acid **(GABA)** and **glycine.**

3. Summation at synapses

a. Spatial summation occurs when two excitatory inputs arrive at a postsynaptic neuron simultaneously. They add together to produce greater depolarization.

  b. Temporal summation occurs when two excitatory inputs arrive at a postsynaptic neuron in rapid succession. Because the resulting postsynaptic depolarizations overlap in time, they add in stepwise fashion.

  c. Facilitation, augmentation, and post-tetanic potentiation occur following tetanic stimulation of the presynaptic neuron. In each of these, depolarization of the postsynaptic neuron is greater than expected because greater than normal amounts of neurotransmitter are released, possibly due to the accumulation of Ca^{2+} in the presynaptic terminal.

   —**Long-term potentiation** (memory) involves new protein synthesis.

 4. Neurotransmitters

  a. ACh (see V B)

  b. Norepinephrine, epinephrine, and dopamine (Figure 1-7)

   (1) Norepinephrine

    —is the primary transmitter for **postganglionic sympathetic neurons**.

    —is synthesized in the nerve terminal (see Figure 1-7) and released into the synapse to **bind with α or β receptors on the postsynaptic membrane**.

    —is removed from the synapse by **reuptake** or is metabolized in the presynaptic terminal by monoamine oxidase **(MAO)** and catechol-O-methyl transferase **(COMT)**. The **metabolites** are:

    (a) 3,4-Dihydroxymandelic acid (DOMA)

    (b) Normetanephrine (NMN)

    (c) 3-Methoxy-4-hydroxyphenylglycol (MOPG)

    (d) 3-Methoxy-4-hydroxymandelic acid (VMA)

   (2) Epinephrine

    —is secreted, along with norepinephrine, from the adrenal medulla.

   (3) Dopamine

    —is prominent in **midbrain** neurons.

    —is released from the hypothalamus and **inhibits prolactin secretion**.

    —is metabolized by MAO and COMT.

    —**D_1 receptors** activate adenylate cylase via a G_s protein.

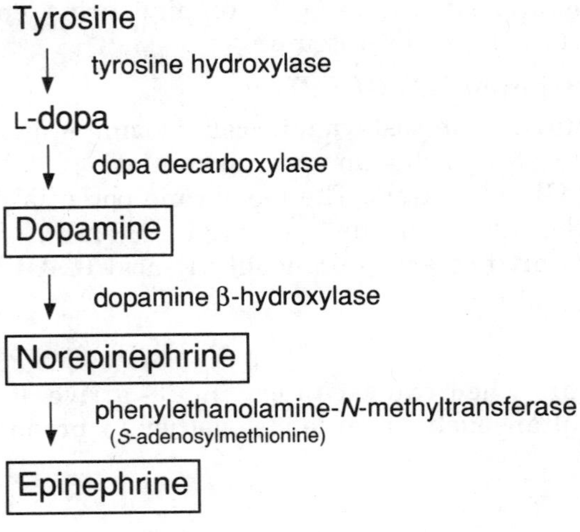

Tyrosine

 ↓ tyrosine hydroxylase

L-dopa

 ↓ dopa decarboxylase

Dopamine

 ↓ dopamine β-hydroxylase

Norepinephrine

 ↓ phenylethanolamine-*N*-methyltransferase
  (*S*-adenosylmethionine)

Epinephrine

Figure 1-7. Synthetic pathway for dopamine, norepinephrine, and epinephrine.

−**D$_2$ receptors** inhibit adenylate cyclase via a G$_i$ protein.
−**Parkinson's disease** involves degeneration of dopaminergic neurons, which utilize the **D$_2$ receptors**.
−**Schizophrenia** involves increased levels of D$_2$ receptors.

c. Serotonin

−is present in high concentrations in the **brainstem**.
−is formed from tryptophan.
−is converted to melatonin in the pineal gland.

d. Histamine

−is formed from histidine.
−is present in neurons of the **hypothalamus**.

e. Glutamate

−is the **most prevalent excitatory neurotransmitter in the brain**.
−The **kainate receptor** is an ion channel for Na$^+$ and K$^+$.

f. GABA

−is an **inhibitory neurotransmitter**.
−is synthesized from glutamate by glutamate decarboxylase.
−Two types of GABA receptors are:
(1) The **GABA$_A$** receptor increases Cl$^-$ conductance and is the site of action of **benzodiazepines** and **barbiturates**.
(2) The **GABA$_B$** receptor increases K$^+$ conductance.

g. Glycine

−is an **inhibitory neurotransmitter** found primarily in the spinal cord and brainstem.
−increases Cl$^-$ conductance.

VI. Skeletal Muscle

A. Muscle structure and filaments (Figure 1-8)

−Each muscle fiber is multinucleate and behaves as a single unit. It contains bundles of **myofibrils,** surrounded by **sarcoplasmic reticulum.**
−Each myofibril contains interdigitating **thick and thin filaments** that are arranged longitudinally in **sarcomeres.** Repeating units of sarcomeres account for the unique banding pattern in striated muscle. A sarcomere runs from **Z line to Z line.**

1. Thick filaments

−contain **myosin.** Each myosin molecule has two "heads" attached to a single tail. The myosin heads bind ATP and actin and are involved in cross-bridge formation.
−are present in the **A band** in the center of the sarcomere.

2. Thin filaments

−contain **actin, tropomyosin,** and **troponin.** Troponin is the **regulatory protein** that permits cross-bridge formation when it binds Ca^{2+}.
−are anchored at the Z lines.
−are present in the **I bands**.
−interdigitate with the thick filaments in a portion of the A band.

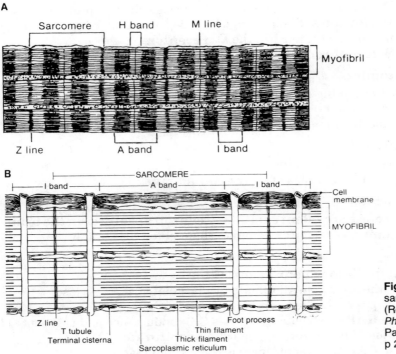

A

Sarcomere   H band   M line

Myofibril

Z line   A band   I band

B

SARCOMERE

I band   A band   I band

Cell membrane

MYOFIBRIL

Z line
T tubule
Terminal cisterna

Foot process
Thin filament
Thick filament
Sarcoplasmic reticulum

Figure 1-8. Structure of the sarcomere in skeletal muscle. (Reprinted from Bullock J et al. *Physiology,* 2nd ed. Malvern, Pa. Williams & Wilkins, 1991, p 29.)

3. T tubules

   —are an extensive tubular network, open to the extracellular space, which carry the depolarization to the cell interior.

   —are located at the junctions of A bands and I bands.

4. Sarcoplasmic reticulum (SR)

   —is the internal tubular structure that is the **site of Ca^{2+} storage and release** for excitation–contraction coupling.

   —has **terminal cisternae** that make intimate contact with the T tubules.

   —membrane contains **Ca^{2+}–ATPase (pumps),** which transport Ca^{2+} from intracellular fluid into the SR interior, keeping intracellular $[Ca^{2+}]$ low.. Within the SR, Ca^{2+} is bound loosely to **calsequestrin,** to be released upon depolarization.

B. Steps in excitation–contraction coupling in skeletal muscle (Figures 1-9 and 1-10)

1. **Action potentials** in the muscle cell membrane initiate depolarization of the T tubules.

2. **Depolarization of the T tubules opens Ca^{2+} channels in the nearby SR,** causing release of Ca^{2+} from the SR into the intracellular fluid.

3. **Intracellular $[Ca^{2+}]$ increases.**

4. **Ca^{2+} binds to troponin C** on the thin filaments, causing a conformational change in troponin.

   a. Tropomyosin is moved out of the way so that the **cross-bridge cycle** can begin.

   b. Actin and myosin bind, the heads of the cross-bridges pivot, the thin and thick filaments slide over each other, and ATP is hydrolyzed.

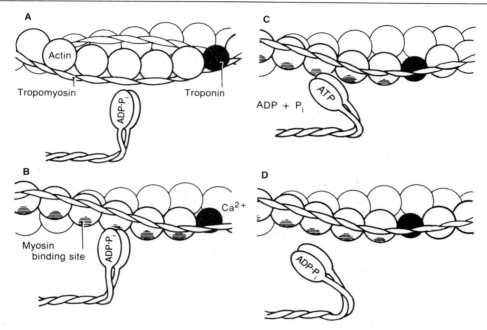

Figure 1-9. The cross-bridge cycle in skeletal muscle. (Reprinted from Bullock J et al. *Physiology,* 2nd ed. Malvern, Pa. Williams & Wilkins, 1991, p 30.)

    c. Subsequently, the cross-bridges break and a new molecule of ATP binds to the myosin head so that a new cycle can begin.

    d. Cross-bridge cycling continues as long as Ca^{2+} is bound to troponin C.

  5. Relaxation occurs when Ca^{2+} uptake into the SR lowers the intracellular $[Ca^{2+}]$. ATP is consumed in the process of Ca^{2+} uptake (as well as during the cross-bridge cycle).

  6. Mechanism of tetanus. A single action potential results in release of a standard amount of Ca^{2+} from the SR and a single twitch (see Figure 1-10). If the muscle is stimulated repeatedly, then more Ca^{2+} is released from the SR, there is a greater rise in intracellular $[Ca^{2+}]$, and there is greater cross-bridge formation. As a result, more tension is generated by the muscle **(tetanus).**

C. Length–tension and force–velocity relationships in muscle

    –Isometric contractions are those that generate force (tension) without shortening.

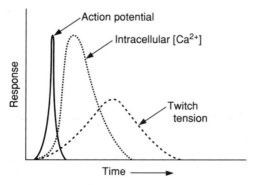

Figure 1-10. Relationship of the action potential, the rise in intracellular $[Ca^{2+}]$, and muscle contraction in skeletal muscle.

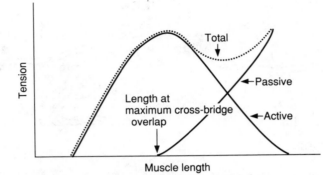

Figure 1-11. Length–tension relationship in skeletal muscle.

1. **Length–tension relationship** (Figure 1-11)

   —measures tension developed during isometric contractions when the muscle is set to fixed lengths **(preload)**.

   a. **Passive tension** is the tension developed by stretching the muscle to different lengths.

   b. **Total tension** is the tension developed when the muscle is stimulated to contract at different lengths.

   c. **Active tension** is the difference between total tension and passive tension.

      —It represents the active force developed from contraction of the muscle, and can be explained by the cross-bridge cycle model. **Tension developed is proportional to the number of cross-bridges formed.** It will be maximum when there is maximum overlap of thick and thin filaments. When the muscle is stretched to greater lengths, the number of cross-bridges is reduced. When muscle length is decreased, the thin filaments collide with each other, also reducing tension.

2. **Force–velocity relationship** (Figure 1-12)

   —measures the velocity of shortening of isotonic contractions when the muscle is challenged with different **afterloads**.

   —The velocity of shortening **decreases as the afterload** (against which the muscle must contract) **increases**.

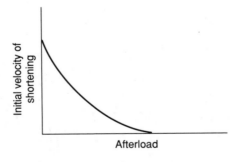

Figure 1-12. Force–velocity relationship in skeletal muscle.

VII. Smooth Muscle

–has thick and thin filaments, but they are not arranged in sarcomeres; therefore, they appear homogeneous rather than striated.

A. Types of smooth muscle

1. **Multi-unit smooth muscle**

   –is present in the **iris, ciliary muscle of the lens,** and **vas deferens.**
   –behaves like separate motor units.
   –has little or no electrical coupling between cells.
   –is **densely innervated** and contraction is under neural control.

2. **Unitary (single-unit) smooth muscle**

   –is the most common type and is present in the **uterus, gastrointestinal tract, ureter,** and **bladder.**
   –is spontaneously active (**slow waves**) and exhibits "pacemaker" activity (see Chapter 6), which is modulated by hormones and neurotransmitters.
   –has a high degree of electrical coupling between cells so that coordinated contraction of the organ can occur (e.g., bladder).

3. **Vascular smooth muscle**

   –has properties of both multi-unit and single-unit smooth muscle.

B. Steps in excitation–contraction coupling in smooth muscle

–The mechanism of excitation–contraction coupling is different from that which occurs in skeletal muscle.
–There is **no troponin;** instead, Ca^{2+} regulates myosin on the thick filaments.

1. **Depolarization of the cell membrane opens voltage-dependent Ca^{2+} channels** and Ca^{2+} flows into the cell down its electrochemical gradient, increasing the intracellular $[Ca^{2+}]$.

2. The Ca^{2+} that enters across the cell membrane may cause release of additional Ca^{2+} from the SR through Ca^{2+}–gated Ca^{2+} channels. Hormones and neurotransmitters also release Ca^{2+} from the SR through **IP_3–gated Ca^{2+} channels.**

3. **Intracellular $[Ca^{2+}]$ increases.**

4. Ca^{2+} binds to calmodulin and the Ca^{2+}–calmodulin complex binds to and **activates myosin light-chain kinase.** When activated, myosin light-chain kinase **phosphorylates myosin** and allows it to bind to actin. Shortening follows.

5. Dephosphorylation of myosin produces relaxation.

Review Test

Directions: Each of the numbered items or incomplete statements in this section is followed by answers or by completions of the statement. Select the **one** lettered answer or completion that is **best** in each case.

1. Which of the following characteristics is shared by simple and facilitated diffusion of glucose?

(A) Occurs down an electrochemical gradient
(B) Is saturable
(C) Requires metabolic energy
(D) Is inhibited by the presence of galactose
(E) Requires a Na^+-gradient

2. Solutions A and B are separated by a semipermeable membrane that is permeable to K^+ but not to Cl^-. Solution A is 100 mM KCl and solution B is 1 mM KCl. Which of the following statements is true about solution A and solution B?

(A) K^+ ions will diffuse from solution A to solution B until the $[K^+]$ of both solutions is 50.5 mM
(B) K^+ ions will diffuse from solution B to solution A until the $[K^+]$ of both solutions is 50.5 mM
(C) KCl will diffuse from solution A to solution B until the [KCl] of both solutions is 50.5 mM
(D) K^+ will diffuse from solution A to solution B until a membrane potential develops with solution A negative with respect to solution B
(E) K^+ will diffuse from solution A to solution B until a membrane potential develops with solution A positive with respect to solution B

3. Which of the following is the correct temporal sequence for events at the neuromuscular junction?

(A) Action potential in the motor nerve; depolarization of the muscle end plate; uptake of Ca^{2+} into the presynaptic nerve terminal
(B) Uptake of Ca^{2+} into the presynaptic terminal; release of ACh; depolarization of the muscle end plate
(C) Release of ACh; action potential in the motor nerve; action potential in the muscle
(D) Uptake of Ca^{2+} into the motor end plate; action potential in the motor end plate; action potential in the muscle
(E) Release of ACh; action potential in the muscle end plate; action potential in the muscle

4. Each of the following is an excitatory neurotransmitter in the central nervous system EXCEPT

(A) norepinephrine
(B) glutamate
(C) GABA
(D) serotonin
(E) histamine

5. Which of the following characteristics or components is shared by skeletal and smooth muscle?

(A) Thick and thin filaments arranged in sarcomeres
(B) Troponin
(C) Elevation of intracellular $[Ca^{2+}]$ for excitation–contraction coupling
(D) Spontaneous depolarization of the membrane potential
(E) High degree of electrical coupling between cells

6. Repeated stimulation of a skeletal muscle fiber causes tetanic contraction because the intracellular concentration of which of the following solutes increases and remains at high levels?

(A) Na^+
(B) K^+
(C) Troponin
(D) ATP
(E) Ca^{2+}

7. A person with myasthenia gravis notes increased muscle strength when he is treated with an AChE inhibitor. The basis for his improvement is increased

(A) amount of ACh released from motor nerves
(B) levels of ACh at the muscle end plates
(C) number of ACh receptors on the muscle end plates
(D) amount of norepinephrine released from motor nerves
(E) synthesis of norepinephrine in motor nerves

8. In error, a patient is infused with large volumes of a solution that causes lysis of his red blood cells. The solution was most likely

(A) isotonic NaCl
(B) isotonic mannitol
(C) hypertonic mannitol
(D) hypotonic urea

9. During a nerve action potential, a stimulus is delivered as indicated by the arrow shown in the following figure. In response to the stimulus, a second action potential

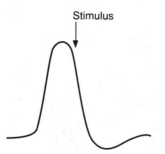

Stimulus

(A) of smaller magnitude will occur
(B) of normal magnitude will occur
(C) of normal magnitude will occur, but it will be delayed in time
(D) will occur, but it will not have an overshoot
(E) will not occur

10. Solution A and solution B are separated by a membrane that is permeable to urea. Solution A is 10 mM urea and solution B is 5 mM urea. If the concentration of urea in solution A is doubled, the flux of urea across the membrane will

(A) double
(B) triple
(C) be unchanged
(D) decrease to one-half
(E) decrease to one-third

11. A muscle cell has an intracellular $[Na^+]$ of 14 mM and an extracellular $[Na^+]$ of 140 mM. Assuming that 2.3 RT/F = 60 mV, what would the membrane potential be if the muscle cell membrane were permeable only to Na^+?

(A) −80 mV
(B) −60 mV
(C) 0 mV
(D) +60 mV
(E) +80 mV

Questions 12–14

The diagram of a nerve action potential applies to Questions 12–14.

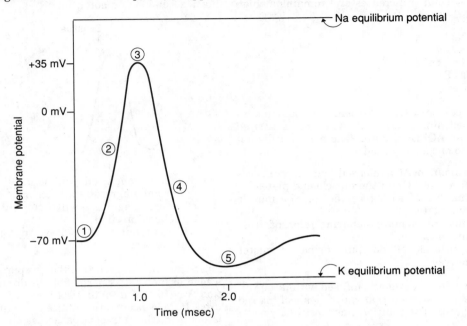

12. At which labeled point on the action potential is the K^+ closest to being at electrochemical equilibrium?

(A) 1
(B) 2
(C) 3
(D) 4
(E) 5

13. What process is responsible for the change in membrane potential occurring between point 1 and point 3?

(A) Movement of Na^+ into the cell
(B) Movement of Na^+ out of the cell
(C) Movement of K^+ into the cell
(D) Movement of K^+ out of the cell
(E) Inhibition of the Na^+–K^+ pump

14. What process is responsible for the change in membrane potential occurring between point 3 and point 4?

(A) Movement of Na^+ into the cell
(B) Movement of Na^+ out of the cell
(C) Movement of K^+ into the cell
(D) Movement of K^+ out of the cell
(E) Inhibition of the Na^+–K^+ pump

15. The rate of conduction of action potentials along a nerve will be increased by

(A) stimulating the Na^+–K^+ pump
(B) inhibiting the Na^+–K^+ pump
(C) decreasing the diameter of the nerve
(D) myelinating the nerve

16. Degeneration of dopaminergic neurons has been implicated in

(A) schizophrenia
(B) Parkinson's disease
(C) myasthenia gravis
(D) curare poisoning

17. The permeability of a solute in a lipid bilayer will be increased by an increase in the

(A) molecular radius
(B) oil/water partition coefficient
(C) thickness of the bilayer
(D) concentration difference of the solute across the bilayer

18. ATP is utilized directly for each of the following processes EXCEPT

(A) accumulation of Ca^{2+} by the sarcoplasmic reticulum
(B) transport of Na^+ from intracellular to extracellular fluid
(C) transport of K^+ from extracellular to intracellular fluid
(D) transport of H^+ from parietal cells into the lumen of the stomach
(E) transport of glucose into muscle cells

19. A drug completely blocks Na^+ channels in nerves. Which of the following effects on the action potential would it be expected to produce?

(A) Block the occurrence of action potentials
(B) Increase the rate of rise of the upstroke of the action potential
(C) Shorten the absolute refractory period
(D) Abolish the hyperpolarizing afterpotential
(E) Increase the Na^+ equilibrium potential

20. At the muscle end plate, ACh causes opening of

(A) Na^+ channels and depolarization toward the Na^+ equilibrium potential
(B) K^+ channels and depolarization toward the K^+ equilibrium potential
(C) Ca^{2+} channels and depolarization toward the Ca^{2+} equilibrium potential
(D) Na^+ and K^+ channels and depolarization to a value halfway between the Na^+ and K^+ equilibrium potentials

21. An inhibitory postsynaptic potential

(A) depolarizes the postsynaptic membrane by opening Na^+ channels
(B) depolarizes the postsynaptic membrane by opening K^+ channels
(C) hyperpolarizes the postsynaptic membrane by opening Ca^{2+} channels
(D) hyperpolarizes the postsynaptic membrane by opening Cl^- channels

22. Which of the following temporal sequences is correct for excitation–contraction coupling in skeletal muscle?

(A) Increased intracellular $[Ca^{2+}]$; action potential in the muscle membrane; crossbridge formation
(B) Action potential in the muscle membrane; depolarization of the T tubules; release of Ca^{2+} from the sarcoplasmic reticulum
(C) Action potential in the muscle membrane; splitting of ATP; binding of Ca^{2+} to troponin C
(D) Release of Ca^{2+} from the sarcoplasmic reticulum; depolarization of the T tubules; binding of Ca^{2+} to troponin C

23. Assuming complete dissociation of all solutes, which of the following solutions would be hyperosmotic to 1 mM NaCl?

(A) 1 mM glucose
(B) 1.5 mM glucose
(C) 1 mM $CaCl_2$
(D) 1 mM sucrose
(E) 1 mM KCl

Directions: The group of items in this section consists of lettered options followed by a set of numbered items. For each item, select the **one** lettered option that is most closely associated with it. Each lettered option may be selected once, more than once, or not at all.

Questions 24–27

Match each numbered description below with the correct type of transport.

(A) Simple diffusion
(B) Facilitated diffusion
(C) Primary active transport
(D) Cotransport
(E) Countertransport

24. Transport of D- and L-glucose proceeds at the same rate down an electrochemical gradient

25. Transport of glucose from the intestinal lumen into a small intestinal cell is inhibited when the usual Na^+ gradient across the membrane is abolished

26. Upstroke of the nerve action potential

27. Transport of K^+ ions from extracellular fluid to intracellular fluid is inhibited by digitalis

Answers and Explanations

1–A. Both types of transport occur "downhill," and do not require metabolic energy. Saturability and inhibition by other sugars is characteristic only of carrier-mediated glucose transport.

2–D. Since the membrane is permeable only to K^+ ions, K^+ will diffuse down its concentration gradient from solution A to solution B, leaving some Cl^- ions behind in solution A. A diffusion potential will be created, with solution A negative with respect to solution B. Generation of a diffusion potential involves movement of only a few ions and so does not cause a change in concentration of the bulk solutions.

3–B. ACh is stored in vesicles and is released when an action potential in the motor nerve opens Ca^{2+} channels in the presynaptic terminal. ACh diffuses across the synaptic cleft and opens Na^+ and K^+ channels in the muscle end plate, depolarizing it (but not producing an action potential). Depolarization of the muscle end plate causes local currents in adjacent muscle membrane, depolarizing it to threshold and producing action potentials.

4–C. GABA is an inhibitory neurotransmitter. Norepinephrine, glutamate, serotonin, and histamine are excitatory neurotransmitters.

5–C. An elevation of intracellular $[Ca^{2+}]$ is common to the mechanism of excitation–contraction coupling in skeletal and smooth muscle. In skeletal muscle, the Ca^{2+} binds troponin C, initiating the cross-bridge cycle. In smooth muscle, Ca^{2+} binds to calmodulin and the Ca^{2+}–calmodulin complex activates myosin light-chain kinase, which phosphorylates myosin so that shortening can occur. The striated appearance of the sarcomeres and the presence of troponin are characteristic of skeletal, not smooth, muscle. Spontaneous depolarizations and gap junctions are characteristics of unitary smooth muscle but not skeletal muscle.

6–E. During repeated stimulation of a muscle fiber, Ca^{2+} is released from the sarcoplasmic reticulum faster than it can be reaccumulated, so the intracellular $[Ca^{2+}]$ does not fall back to resting levels as it would after a single twitch. The increased $[Ca^{2+}]$ allows more cross-bridges to form and therefore increased tension (tetanus). Intracellular Na^+ and K^+ concentrations do not change during the action potential; very few Na^+ or K^+ ions move into or out of the muscle cell, and so bulk concentrations are unaffected. ATP levels would, if anything, fall during tetanus because it is consumed during cross-bridge formation.

7–B. Myasthenia gravis is characterized by a decreased density of ACh receptors at the muscle end plate. An AChE inhibitor blocks degradation of ACh in the neuromuscular junction, so that levels at the muscle end plate will remain high, partially compensating for the deficiency of receptors.

8–D. Lysis of the patient's red blood cells was caused by entry of water and swelling of the cells to the point of rupture. Water flow into the red cells would occur if the extracellular fluid was hypotonic (had a lower osmotic pressure) to the intracellular fluid—hypotonic urea. By definition, isotonic solutions do not cause flow of water into or out of cells because osmotic pressure is the same on both sides of the cell membrane. Hypertonic mannitol would cause shrinkage of the red cells.

9–E. Because the stimulus was delivered during the absolute refractory period, no action potential will result. The inactivation gates of the Na^+ channel have been closed by depolarization and remain closed until the membrane has repolarized. As long as the inactivation gates are closed, the Na^+ channels cannot be opened for another action potential to occur.

10–B. Flux is proportional to the concentration difference across the membrane, $J = PA(C_A - C_B)$. Originally, $C_A - C_B = 10 \text{ mM} - 5 \text{ mM} = 5 \text{ mM}$. When the urea concentration was doubled in solution A, the concentration difference became $20 \text{ mM} - 5 \text{ mM} = 15 \text{ mM}$, or three times the original. Therefore, the flux would also triple.

11–D. The Nernst equation is used to calculate the equilibrium potential for a single ion. In applying the Nernst equation, we assume that the membrane is freely permeable to that ion alone. $E_{Na^+} = 2.3 \text{ RT}/zF \log C_e/C_i = 60 \text{ mV} \log 140/14 = 60 \text{ mV} \log 10 = 60 \text{ mV}$. Notice that signs were ignored and the higher concentration was simply put in the numerator to make the log calculation easier. To determine whether E_{Na^+} is $+60 \text{ mV}$ or -60 mV, use the intuitive approach—Na^+ will diffuse from extracellular to intracellular fluid down its concentration gradient, making the cell interior positive.

12–E. The hyperpolarizing afterpotential represents the period during which K^+ permeability is highest, and the membrane potential is closest to the K^+ equilibrium potential. At that point, K^+ is closest to being at electrochemical equilibrium—the force driving K^+ movement out of the cell down its chemical gradient is balanced by the force driving K^+ into the cell down its electrical gradient.

13–A. The upstroke of the nerve action potential is caused by opening of the Na^+ channels (once the membrane is depolarized to threshold). When the Na^+ channels open, Na^+ moves into the cell down its electrochemical gradient, driving the membrane potential toward the Na^+ equilibrium potential.

14–D. The process responsible for repolarization is the opening of K^+ channels. The K^+ permeability becomes very high and drives the membrane potential toward the K^+ equilibrium potential by flow of K^+ out of the cell.

15–D. Myelin insulates the nerve, thereby increasing conduction velocity; action potentials can be generated only at the nodes of Ranvier where there are breaks in the insulation. Activity of the Na^+–K^+ pump does not have a direct effect on formation or conduction of action potentials. Decreasing nerve diameter would increase internal resistance and therefore slow the conduction velocity.

16–B. Dopaminergic neurons and D_2 receptors are deficient in people with Parkinson's disease. Schizophrenia involves increased levels of D_2 receptors. Myasthenia gravis and curare poisoning involve the neuromuscular junction, which uses ACh as a neurotransmitter.

17–B. Increasing oil/water partition coefficient increases solubility in a lipid bilayer and so increases permeability. Increasing molecular radius and increased thickness will decrease permeability. The concentration difference of the solute has no bearing on permeability.

18–E. All of the processes listed are examples of primary active transport (and therefore use ATP directly) except for glucose uptake into muscle cells, which occurs by facilitated diffusion.

19–A. Complete blockade of the Na^+ channels would prevent action potentials. The upstroke of the action potential depends on Na^+ entry into the cell through these channels and so would also be abolished. The absolute refractory period would be lengthened because it is based on availability of the Na^+ channels. The hyperpolarizing afterpotential is related to increased K^+ permeability. The Na^+ equilibrium potential is calculated from the Nernst equation and is the theoretical potential at electrochemical equilibrium (and does not depend on whether the Na^+ channels are open or closed).

20–D. Binding of ACh to receptors in the muscle end plate opens channels that allow passage of both Na^+ and K^+ ions. Na^+ ions will flow into the cell down its electrochemical gradient, and K^+ ions will flow out of the cell down its electrochemical gradient. The resulting membrane potential will be about halfway between their respective equilibrium potentials.

21–D. An inhibitory postsynaptic potential hyperpolarizes the postsynaptic membrane, taking it farther from threshold. Opening Cl^- channels would hyperpolarize by driving the membrane potential toward the Cl^- equilibrium potential (about -90 mV). Opening Ca^{2+} channels would depolarize the postsynaptic membrane by driving it toward the Ca^{2+} equilibrium potential.

22–B. The correct sequence is action potential in the muscle membrane; depolarization of the T tubules; release of Ca^{2+} from the sarcoplasmic reticulum; binding of Ca^{2+} to troponin C; cross-bridge formation; splitting of ATP.

23–C. Osmolarity is the concentration of particles (osmolarity $= g \times C$). When two solutions are compared, that with the higher osmolarity is hyperosmotic. The 1 mM $CaCl_2$ solution (osmolarity $= 3$ mOsmol) is hyperosmotic to 1 mM NaCl (osmolarity $= 2$ mOsmol). The 1 mM glucose, 1.5 mM glucose, and 1 mM sucrose solutions are hyposmotic to 1 mM NaCl, while 1 mM KCl is isosmotic.

24–A. Only two types of transport occur "downhill"—simple and facilitated diffusion. Since there is no stereospecificity for the D- or L-isomer, one can conclude that the transport is not carrier-mediated and so must be simple diffusion.

25–D. The "usual" Na^+ gradient is such that the $[Na^+]$ is higher in extracellular than in intra-cellular fluid (maintained by the Na^+–K^+ pump). Two forms of transport are energized by such a Na^+ gradient—cotransport and countertransport. Since glucose is moving in the same direction as Na^+, one can conclude that it is cotransport.

26–A. Depolarization opens Na^+ channels in the nerve membrane, allowing Na^+ ions to diffuse into the cell down an electrochemical gradient. This is a purely passive process.

27–C. Transport of K^+ from extracellular fluid to intracellular fluid occurs against an electro-chemical gradient and is an example of primary active transport. The carrier is the Na^+–K^+ pump, which is inhibited by cardiac glycosides such as digitalis.

2

Neurophysiology

I. Autonomic Nervous System

- —is a set of **efferent pathways from the central nervous system** (CNS) that innervates and regulates **smooth muscle, cardiac muscle,** and **glands**.
- —is distinct from the **somatic nervous system**, which innervates **skeletal muscle**.
- —has three divisions: **sympathetic, parasympathetic,** and **enteric** (the enteric division is discussed in Chapter 6).

   A. Organization of the autonomic nervous system (Table 2-1 and Figure 2-1)

Table 2-1. Organization of the Autonomic Nervous System

Characteristic	Sympathetic	Parasympathetic	Somatic*
Origin of preganglionic nerve	Nuclei of spinal cord segments T1–T12; L1–L3 (thoracolumbar)	Nuclei of cranial nerves III, VII, IX and X; spinal cord segments S2–S4 (craniosacral)	
Length of preganglionic nerve axon	Short	Long	
Neurotransmitter in ganglion	ACh	ACh	
Receptor type in ganglion	Nicotinic	Nicotinic	
Length of postganglionic nerve axon	Long	Short	
Effector organ	Smooth and cardiac muscle; glands	Smooth and cardiac muscle; glands	Skeletal muscle
Neurotransmitter in effector organ	Norepinephrine (except sweat glands, which use ACh)	ACh	ACh (synapse is neuromuscular junction)
Receptor type in effector organ	α_1, α_2, β_1, and β_2	Muscarinic	Nicotinic

*Somatic nervous system has been included for comparison.

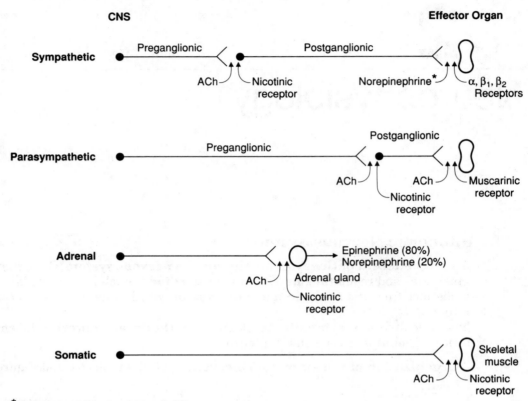

*Except sweat glands, which use ACh.

Figure 2-1. Organization of the autonomic nervous system.

1. **Synapses between neurons** are made in the **autonomic ganglia**.

    a. **Parasympathetic ganglia** are located in the effector organs.

    b. **Sympathetic ganglia** are located in paravertebral chains.

2. **Preganglionic neurons** have their cell bodies in the CNS and synapse in autonomic ganglia.

3. **Postganglionic neurons** have their cell bodies in the autonomic ganglia and synapse on effector organs (e.g., heart, blood vessels, sweat glands).

4. **Adrenal medulla** is a specialized case. Preganglionic fibers synapse directly on **chromaffin cells** in the adrenal medulla; these cells secrete epinephrine (80%) and norepinephrine (20%) into the circulation (see Figure 2-1).

B. **Receptor types in the autonomic nervous system**

1. **Adrenergic receptors (adrenoreceptors)**

    a. α_1 **Receptors**

    −are **located on smooth muscle** (except bronchial smooth muscle).

    −produce **excitation**.

–are equally sensitive to norepinephrine and epinephrine, but only norepinephrine is present in concentrations that are high enough to activate α receptors.

–**Mechanism of action**: formation of inositol triphosphate (IP_3) and increase in intracellular $[Ca^{2+}]$.

b. α$_2$ Receptors

–are located in **presynaptic nerve terminals, platelets, fat cells,** and **smooth muscle.**

–often produce inhibition.

–**Mechanism of action**: inhibition of adenylate cyclase and decrease in cyclic adenosine monophosphate (AMP).

c. β$_1$ Receptors

–are located in the **heart.**

–produce **excitation.**

–are sensitive to both norepinephrine and epinephrine, and are **more sensitive than the α receptors.**

–**Mechanism of action**: activation of adenylate cyclase and production of cyclic AMP.

d. β$_2$ Receptors

–are located on **vascular smooth muscle, bronchial smooth muscle,** and in the **gastrointestinal tract.**

–produce **relaxation.**

–are **more sensitive to epinephrine than to norepinephrine.**

–are **more sensitive to epinephrine than is the α receptor.**
For example, when small amounts of epinephrine are released from the adrenal medulla, **vasodilation (β$_2$)** occurs; when larger amounts of epinephrine are released from the adrenal medulla, **vasoconstriction (α)** occurs.

–**Mechanism of action**: see β$_1$ receptors.

2. Cholinergic receptors (cholinoreceptors)

a. Nicotinic receptors

–are located in the **autonomic ganglia** and at the **neuromuscular junction.** The receptors at the two locations are similar but not identical.

–are **activated by acetylcholine (ACh) or nicotine.**

–produce **excitation.**

–**Ganglionic blockers (e.g., hexamethonium, trimethaphan)** block the nicotinic receptor for ACh in the autonomic ganglia but not at the neuromuscular junction.

–**Mechanism of action**: nicotinic ACh receptors are ion channels for Na^+ and K^+.

b. Muscarinic receptors

–are located in the **heart, smooth muscle (except vascular smooth muscle),** and **glands.**

–are **activated by ACh or muscarine.**

–are **inhibitory in the heart** and **excitatory in smooth muscle and glands.**

—**Atropine** blocks the muscarinic receptors for ACh.

—**Mechanism of action:**

(1) **Heart SA node:** inhibition of adenylate cyclase and opening of K^+ channels, which slows the rate of spontaneous depolarization and decreases the heart rate.

(2) **Smooth muscle and glands:** formation of IP_3 and increase in intracellular $[Ca^{2+}]$.

3. **Drugs acting on the autonomic nervous system** (Table 2-2)

C. **Effects of the autonomic nervous system on various organ systems** (Table 2-3)

II. Sensory Systems

A. **Sensory receptors—general**

—are neurons that **transduce environmental signals** into neural signals.

—The environmental signals that can be detected include **mechanical force, light, sound, chemicals,** and **temperature**.

1. **Types of sensory transducers**

a. **Mechanoreceptors**

—Pacinian corpuscles

—Joints

—Stretch receptors in muscle

—Hair cells in auditory and vestibular systems

b. **Photoreceptors**

—Rods and cones of the retina

c. **Chemoreceptors**

—Olfactory receptors

Table 2-2. Prototypes of Drugs that Affect Autonomic Activity

Type of Receptor	Agonist	Antagonist
Adrenergic		
α_1	Norepinephrine Phenylephrine	Phenoxybenzamine Phentolamine Prazosin
α_2	Clonidine	Yohimbine
β_1	Norepinephrine Isoproterenol Dobutamine	Propranolol Metoprolol
β_2	Isoproterenol Albuterol	Propranolol Butoxamine
Cholinergic		
Nicotinic	ACh Nicotine Carbachol	Curare Hexamethonium (ganglion, but not neuromuscular junction)
Muscarinic	ACh Muscarine Carbachol	Atropine

Table 2-3. Effect of the Autonomic Nervous System on Organ Systems

Organ	Sympathetic Action	Sympathetic Receptor	Parasympathetic Action (Receptors are muscarinic)
Heart	↑ heart rate ↑ contractility	β_1 β_1	↓ heart rate ↓ contractility (atria)
Vascular smooth muscle	Constricts blood vessels in skin; splanchnic Dilates blood vessels in skeletal muscle	α β_2	
Gastrointestinal tract	↓ motility Constricts sphincters	α_2, β_2 α_1	↑ motility Relaxes sphincters
Bronchioles	Dilates bronchiolar smooth muscle	β_2	Constricts bronchiolar smooth muscle
Male sex organs	Ejaculation	α	Erection
Bladder	Relaxes bladder wall Constricts sphincter	β_2 α_1	Contracts bladder wall Relaxes sphincter
Sweat glands	↑ sweating	Muscarinic (sympathetic cholinergic)	
Kidney	↑ renin secretion	β_1	
Fat cells	↑ lipolysis ↓ lipolysis	β_1 α_2	

- —Taste receptors
- —Osmoreceptors
- —Carotid body receptors

d. Extremes of temperature and pain

- —Nociceptors

2. Fiber types and conduction velocity (Table 2-4)

3. Receptive field

- —is the region that contains the sensory transducers.
- —can be **excitatory** or **inhibitory**.

4. Steps in sensory transduction

a. Stimulus arrives at the sensory receptor. The stimulus may be a photon of light on the retina, a molecule of NaCl on the tongue, a depression of the skin, and so forth.

b. Ion channels are opened in the sensory receptor, allowing current to flow. Usually the current is inward, which produces **depolarization** of the receptor. The **exception is in the photoreceptor, where light causes hyperpolarization.**

c. The change in membrane potential produced by the stimulus is called the **receptor potential** or **generator potential** (Figure 2-2).

Table 2-4. Characteristics of Nerve Fiber Types

General Fiber Type and Example	Sensory Fiber Type and Example	Diameter	Conduction Velocity
A-alpha Large α-motoneurons	**Ia** Muscle spindle afferents	Largest	Fastest
	Ib Golgi tendon organs	Largest	Fastest
A-beta Touch, pressure	**II** Secondary afferents of muscle spindles; touch and pressure	Medium	Medium
A-gamma γ-Motoneurons to muscle spindles (intrafusal fibers)		Medium	Medium
A-delta Touch, pressure, temperature, and pain	**III** Touch, pressure, fast pain, and temperature	Small	Medium
B Preganglionic autonomic fibers		Small	Medium
C Slow pain; postganglionic autonomic fibers	**IV** Pain and temperature (unmyelinated)	Smallest	Slowest

- —If the receptor potential is depolarizing, it brings the membrane potential closer to threshold.
- —Receptor potentials are **graded in size** depending on the size of the stimulus. If the receptor potential is large enough, the membrane potential will exceed threshold and an action potential will be fired by the sensory neuron.

5. Adaptation of sensory receptors

    a. Slowly adapting receptors (muscle spindle; pressure, slow pain) respond repetitively to a prolonged stimulus.

    b. Rapidly adapting receptors (pacinian corpuscle; light touch) show a decline in action potential frequency with time in response to a constant stimulus.

6. Sensory pathways from the sensory receptor to the cerebral cortex

    a. First-order neurons are the **primary afferent neurons** that re-

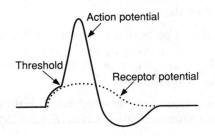

Figure 2-2. Receptor (generator) potential and how it may lead to an action potential.

ceive signals (e.g., light), transduce it, and send the information to the CNS. Cell bodies of the primary afferent neurons are in **dorsal root** or **cranial nerve ganglia**.

b. **Second-order neurons** are located in the spinal cord or brainstem. They receive information from one or more primary afferent neurons and transmit it to the **thalamus**. Axons of second-order neurons usually **cross the midline** at a relay nucleus in the spinal cord before ascending. Therefore, sensory **information originating on one side of the body ascends to the contralateral thalamus**.

c. **Third-order neurons** are located in a sensory nucleus of the thalamus. From here, encoded sensory information ascends to the cerebral cortex.

d. **Fourth-order neurons** are located in the appropriate sensory area of the cortex. The information received results in a conscious **perception** of the stimulus.

B. **Somatosensory system and pain**

—includes the sensations of touch, movement, temperature, and pain.

1. **Pathways in the somatosensory system**

a. **Dorsal column system**

—consists primarily of **group II fibers**, which enter the dorsal root and ascend ipsilaterally to the nuclei gracilis and cuneatus of the medulla. From there, the second-order neurons cross the midline and ascend to the contralateral thalamus.

—processes sensations of **touch, pressure, vibration,** and **movement**.

b. **Anterolateral system**

—consists primarily of **group III and IV fibers**, which enter the spinal cord and terminate in the dorsal horn. Secondary neurons project across the midline to the anterolateral quadrant of the spinal cord and ascend to the contralateral thalamus.

—processes sensations of **temperature** and **pain**.

2. **Mechanoreceptors for touch and pressure** (Table 2-5)

3. **Thalamus**

—Information from different parts of the body is arranged somatotopically.

Table 2-5. Types of Mechanoreceptors

Type of Mechanoreceptor	Description	Sensation Encoded	Adaptation
Pacinian corpuscle	Onion-like structures in the subcutaneous skin (surrounding unmyelinated nerve endings)	Vibration; tapping	Rapidly adapting
Meissner's corpuscle	Present in non-hairy skin	Velocity	Rapidly adapting
Ruffini's corpuscle	Encapsulated	Pressure	Slowly adapting
Merkel's disk	Transducer is on epithelial cells	Location	Slowly adapting

–**Destruction of thalamic nuclei** results in loss of sensation on the contralateral side of the body.

4. Somatosensory cortex—the sensory homunculus

–The major somatosensory areas of the cortex are **SI** and **SII**.
–SI has a somatotopic representation similar to that in the thalamus.
–This "map" of the body is called the **sensory homunculus**.
–The largest areas represent the **face, hands,** and **fingers**, where precise localization is most important.

5. Pain

–is associated with the detection and perception of noxious stimuli (**nociception**).
–The receptors for pain are **free nerve endings** in the skin, muscle, and viscera.
–Neurotransmitters for nociceptors include **substance P**. Inhibition of release of substance P is the basis of pain relief by **opioids**.

a. Fibers for fast pain and slow pain

–**Fast pain** is carried by group III fibers. It has a rapid onset and offset, and is localized.
–**Slow pain** is carried by C fibers. It is characterized as aching, burning, or throbbing that is poorly localized.

b. Referred pain

–Pain of visceral origin is referred to sites on the skin and follows the **dermatome rule**. These sites are innervated by nerves arising from the same segment of the spinal cord.
–**For example**, ischemic heart pain is referred to the chest and shoulder.

C. Vision

1. Optics

a. Refractive power of a lens

–is measured in **diopters**.
–equals the reciprocal of the focal distance in meters.
–**Example:** 10 diopters = 1/10 meters = 10 cm

b. Refractive errors

(1) **Emmetropia—normal.** Light focuses on the retina.
(2) **Hypertropia—far sighted**. Light focuses behind the retina and is corrected with a **convex lens**.
(3) **Myopia—near sighted.** Light focuses in front of the retina and is corrected with a **biconcave lens**.
(4) **Astigmatism**. Curvature of the lens is not uniform and is corrected with a **cylindric lens**.
(5) **Presbyopia** is a result of loss of accommodation power of the lens with the aging process. The **near point** (closest point one can focus by accommodation of lens) moves farther from the eye and is corrected with a **convex lens**.

2. Layers of the retina (Figure 2-3)

a. Pigment cells contain melanin and absorb stray light.

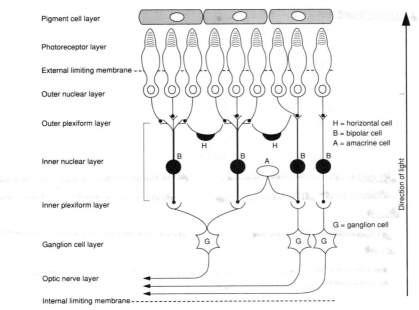

Figure 2-3. Cellular layers of the retina. (Reprinted from Bullock J et al: *Physiology*, 3rd ed. Malvern, Pa, Williams & Wilkins, 1995, p 76.)

    b. Receptor cells are rods and cones (Table 2-6). Because rods and cones are not present on the optic disk, a **blind spot** is created.

    c. Bipolar cells and ganglion cells. The bipolar cells synapse on the receptor cells (rods and cones) and terminate on the ganglion cells.

        (1) Few cones synapse on a single bipolar cell, which synapses on a single ganglion cell. This arrangement is the basis for the high acuity and low sensitivity of the cones. In the fovea where acuity is highest, the ratio of cones to bipolar cells is 1:1.

        (2) Many rods synapse on a single bipolar cell; therefore, there is less acuity. There is also greater sensitivity because light striking any one of the rods will activate the bipolar cell.

    d. Horizontal cells and amacrine cells form local circuits with the bipolar cells.

Table 2-6. Functions of Rods and Cones

Function	Rods	Cones
Sensitivity to light	Sensitive to low-intensity light; night vision	Sensitive to high-intensity light; day vision
Acuity	Lower visual acuity Not present in fovea	Higher visual acuity Present in fovea
Dark adaptation	Rods adapt later	Cones adapt first
Color vision	No	Yes

3. Optic pathways and lesions (Figure 2-4)

- —Axons of the ganglion cells form the optic nerve and optic tract, ending in the lateral geniculate body of the thalamus.
- —The fibers from each **nasal hemiretina** cross **at the optic chiasm**, while the fibers from each temporal hemiretina remain ipsilateral. Therefore, fibers from the **left nasal hemiretina** and fibers from the **right temporal hemiretina** will form the **right optic tract** and synapse on the right lateral geniculate body.
- —Fibers from the lateral geniculate body form the **geniculocalcarine tract** and pass to the **occipital lobe of the cortex**.

a. Cutting the optic nerve causes blindness in the ipsilateral eye.

b. Cutting the optic chiasm causes heteronymous bitemporal hemianopia.

c. Cutting the optic tract causes homonymous contralateral hemianopia.

d. Cutting the geniculocalcarine tract causes homonymous hemianopia with **macular sparing**.

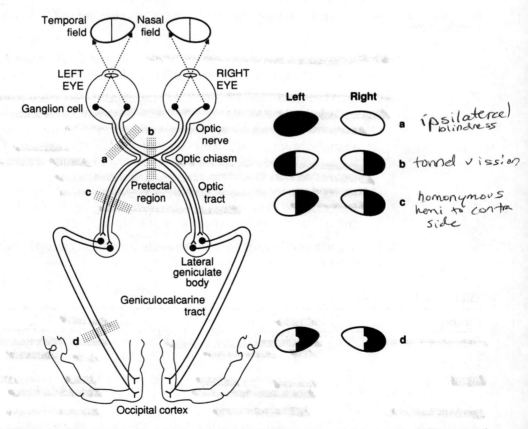

Figure 2-4. Effects of lesions at various levels of the optic pathway. (Modified from Ganong WF: *Review of Medical Physiology*, 16th ed. Norwalk, Ct, Appleton & Lange, 1993, p 135.)

4. Steps in photoreception in the rods

  −The photosensitive element is **rhodopsin**, which is composed of **scotopsin** (a protein) and **retinene$_1$** (an aldehyde of vitamin A).

  a. Light on the retina converts **11-*cis* rhodopsin to all-*trans* rhodopsin**. A series of intermediates is then formed, one of which is ***meta*-rhodopsin II**.

   −**Vitamin A** is necessary for the regeneration of rhodopsin. A deficiency of vitamin A causes **night blindness**.

  b. *Meta*-rhodopsin II **activates a G protein** called **transducin**, which in turn activates a phosphodiesterase.

  c. Phosphodiesterase catalyzes the conversion of cyclic guanosine monophosphate (GMP) to 5′-GMP, and **cyclic GMP levels decrease**.

  d. Decreased levels of cyclic GMP cause **closure of Na$^+$ channels** and therefore **hyperpolarization** of the receptor cell membrane. Increasing light intensity increases the degree of hyperpolarization.

  e. When the receptor cell is hyperpolarized, **release of either an excitatory neurotransmitter or an inhibitory neurotransmitter decreases**.

   (1) If the neurotransmitter is excitatory, then the response of the bipolar or horizontal cell to light is hyperpolarization.

   (2) If the neurotransmitter is inhibitory, then the response of the bipolar or horizontal cell to light is depolarization.

5. Receptive visual fields

  a. Receptive fields of the ganglion cells and lateral geniculate cells

   (1) Each bipolar cell receives input from many receptor cells. In turn, each ganglion cell receives input from many bipolar cells. The receptor cells connected to a ganglion cell form the **center of its receptor field**. The receptor cells connected to ganglion cells via horizontal cells form the **surround of its receptive field**. (Remember that the response of bipolar and horizontal cells to light depends on whether that cell releases an excitatory or inhibitory neurotransmitter.)

   (2) On-center, off-surround is one pattern of ganglion cell receptive field. Light hitting the center of the receptive field depolarizes (excites) the ganglion cell, while light hitting the surround of the receptive field hyperpolarizes (inhibits) the ganglion cell. **Off-center, on-surround** is another possible pattern.

   (3) Lateral geniculate cells of the thalamus retain the on-center, off-surround or off-center, on-surround pattern transmitted from the ganglion cell.

  b. Receptive fields of the visual cortex

   −Neurons in the visual cortex detect shape and orientation of figures.

   −Three cell types are responsible:

   (1) Simple cells have similar center-surround, off-on patterns, but are elongated rods rather than concentric circles. They respond best to **bars of light that have the correct position and orientation**.

(2) Complex cells respond best to **moving bars** or **edges of light** with the correct orientation.

(3) Hypercomplex cells respond best to **lines with particular length** and to **curves and angles**.

D. Audition

1. Sound waves
  —**Frequency** is measured in **hertz** (Hz).
  —**Intensity** is measured in **decibels**, a log scale.

2. Structure of the ear

a. Outer ear
  —directs the sound waves into the auditory canal.

b. Middle ear
  —is air-filled.
  —contains the **tympanic membrane** and the **auditory ossicles** (malleus, incus, and stapes). The stapes inserts into the **oval window**, a membrane between the middle ear and the inner ear.
  —Sound waves cause the tympanic membrane to vibrate. In turn, the **ossicles vibrate**, pushing the stapes into the oval window and **displacing fluid** in the **inner ear** (see II D 2 c).
  —**Sound energy is amplified** by the lever action of the ossicles and the concentration of sound waves from the large tympanic membrane onto the small oval window.

c. Inner ear (Figure 2-5)
  —is fluid-filled.
  —consists of a bony labyrinth (**semicircular canals, cochlea,** and **vestibule**) and a series of ducts called the **membranous labyrinth**. The fluid outside the ducts is **perilymph**; the fluid inside the ducts is **endolymph**.
  (1) Structure of the cochlea: three tubular canals
    (a) The **scala vestibuli** and **scala tympani** contain **perilymph**, which has a **high [Na^+]**.
    (b) The **scala media** contains **endolymph**, which has a **high [K^+]**. The scala media is bordered by the **basilar membrane**, the site of the **organ of Corti**.
  (2) Structure of the organ of Corti
    —is located on the basilar membrane.
    —contains the **receptor cells** (inner hair cells and outer hair cells) for auditory stimuli. **Cilia** protrude from the hair cells.
    —**Inner hair cells** are arranged in single rows and are **few** in number.
    —**Outer hair cells** are arranged in parallel rows and are **greater in number** than the inner hair cells.
    —The **spiral ganglion** contains the cell bodies of the auditory nerve (cranial nerve [CN] VIII), which synapse on the hair cells.

3. Steps in auditory transduction by the organ of Corti (see Figure 2-5)
  —The cilia on the hair cells are imbedded in the **tectorial membrane**.

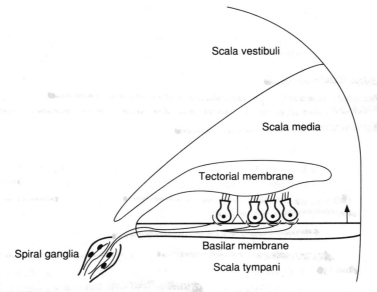

Figure 2-5. Organ of Corti and auditory transduction.

a. Sound waves cause **vibration** of the organ of Corti. Because the basilar membrane is more elastic than the tectorial membrane, vibration of the basilar membrane causes the hair cells to bend by a shearing force as they push against the tectorial membrane.

b. **Bending of the cilia** causes changes in K^+ conductance of the hair cell membrane. Bending in one direction causes depolarization; bending in the other direction causes hyperpolarization. The oscillating potential that results is called the **cochlear microphonic potential**.

c. The oscillating potential of the hair cells causes intermittent firing of the cochlear nerves.

4. How sound is encoded

–The frequency that activates a particular hair cell depends on the location of the hair cell along the basilar membrane.

a. The **base of the basilar membrane** (near the oval and round windows) is narrow and stiff. It responds best to **high frequencies**.

b. The **apex of the basilar membrane** (near the helicotrema) is wide and compliant. It responds best to **low frequencies**.

5. Central auditory pathways

–Fibers ascend through the **lateral lemniscus** to the **inferior colliculus** to the **medial geniculate nucleus** of the thalamus to the **auditory cortex**.

–Fibers may be **crossed or uncrossed**, resulting in a mixture of ascending auditory fibers representing both ears at all higher levels. Therefore, lesions of the cochlea of one ear cause unilateral deafness, but more central unilateral lesions do not.

—There is **tonotopic representation** of frequencies at all levels of the central auditory pathway.

—Complex feature discrimination (such as the ability to recognize a patterned sequence) is a property of the cerebral cortex.

E. Vestibular system

—detects angular and linear acceleration of the head.

—Reflex adjustments of the head, eyes, and postural muscles provide a stable visual image and steady posture.

1. Structure of the vestibular organ

a. It is a membranous labyrinth consisting of **three perpendicular semicircular canals, a utricle,** and a **saccule**. The semicircular canals detect angular acceleration or rotation. The utricle and saccule detect linear acceleration.

b. The canals are **filled with endolymph** and are bathed in perilymph.

c. The **receptors are hair cells** located at the end of each semicircular canal. Cilia on the hair cells are imbedded in a gelatinous structure called the **cupula**. A single long cilium is called the **kinocilium**; smaller cilia are called **stereocilia** (Figure 2-6).

2. Steps in vestibular transduction—angular acceleration (see Figure 2-6)

a. During a **counterclockwise (left) rotation of the head**, the horizontal semicircular canal and its attached cupula initially move more

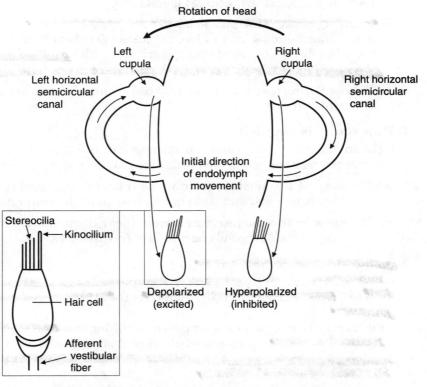

Figure 2-6. The semicircular canals and vestibular transduction.

quickly than the endolymph fluid. The cupula is dragged through the endolymph, causing bending of the cilia on the hair cells.

    b. If the **stereocilia are bent toward the kinocilium, the hair cell depolarizes** (excitation). If the **stereocilia are bent away from the kinocilium, the hair cell hyperpolarizes** (inhibition). Therefore, during the initial counterclockwise (left) rotation, the left horizontal canal is excited and the right horizontal canal is inhibited.

    c. After several seconds, the endolymph "catches up" with the movement of the head and the cupula, and the cilia return to their upright position and are no longer depolarized or hyperpolarized.

    d. When the head suddenly stops moving, the endolymph continues to move counterclockwise (left) for awhile, dragging the cilia in the opposite direction. Therefore, if the hair cell was depolarized with the initial rotation, it will now hyperpolarize. If it was hyperpolarized initially, it will now depolarize. Therefore, when the head stops moving, the left horizontal canal will be inhibited and the right canal will be excited.

  3. Vestibular–ocular reflexes

    a. Nystagmus

      —An initial rotation of the head causes the eyes to move slowly in the opposite direction to maintain visual fixation. When the limit of eye movement is reached, the eyes rapidly snap back, then move slowly again.

      —The **direction of the nystagmus** is defined as the direction of the **fast (rapid eye) movement** and, therefore, occurs in the **same direction as the head rotation**.

    b. Post-rotatory nystagmus

      —occurs in the **opposite direction of the head rotation**.

F. Olfaction

  1. Olfactory pathway

    a. Receptor cells

      —are located in the olfactory epithelium.

      —are **true neurons** that conduct action potentials into the CNS.

      —Basal cells of the olfactory epithelium are undifferentiated stem cells that **continuously turn over** and replace the olfactory receptor cells (neurons). These are the **only examples** in the adult human where neurons replace themselves.

    b. CN I (olfactory)

      —carries information from olfactory receptor cells to the olfactory bulb.

      —The axons of the olfactory nerves are **unmyelinated C fibers** and are among the **smallest** (and therefore **slowest**) in the nervous system.

      —Olfactory epithelium is also innervated by **CN V** (trigeminal), which detects **noxious or painful stimuli** such as ammonia.

      —The olfactory nerves pass through the **cribriform plate** on their way to the olfactory bulb. **Fractures of the cribriform plate** sever input to the olfactory bulb and reduce (**hyposmia**) or eliminate (**anosmia**) the sense of smell. Responses to ammonia will be intact following fracture of the cribriform plate since these responses are carried on CN V.

    c. Mitral cells in the olfactory bulb

      —are the second-order neurons.

      —Output of the mitral cells forms the olfactory tract, which projects to the **prepiriform cortex**.

  2. Steps in transduction in the olfactory receptor neurons

    a. Odorant molecules bind to receptors on the cilia of the olfactory receptor neurons.

    b. When the receptors are activated, they **activate G proteins**, which in turn activate adenylate cyclase.

    c. The **increase in intracellular [cyclic AMP]** opens Na^+ channels in the olfactory receptor membrane, producing a **depolarizing receptor potential**.

    d. The receptor potential depolarizes the initial segment of the axon to threshold, and **action potentials** are generated and propagated.

G. Taste

  1. Taste pathways

    a. Receptor cells for taste sensation are located in the taste buds. These cells are covered with microvilli, which increases the surface area for taste chemicals to bind. In contrast to olfactory receptor cells, **taste receptors are not neurons**. The taste buds are located on **specialized papillae**.

    b. The **anterior two-thirds of the tongue** has **fungiform papillae**. It detects **salty and sweet** sensations, and is innervated by **CN VII** (chorda tympani).

    c. The **posterior one-third of the tongue** has **circumvallate and foliate papillae**. It detects sour and bitter sensations, and is innervated by **CN IX** (glossopharyngeal). The back of the throat and epiglottis is innervated by **CN X**.

    d. CNN VII, IX, and X enter the medulla, ascend in the **solitary tract**, and terminate on second-order taste neurons in the **solitary nucleus**. They project, primarily ipsilaterally, to the ventral posteromedial nucleus of the thalamus and finally to the taste cortex.

  2. Steps in taste transduction

    —**Taste chemicals** (sour, sweet, salt, and bitter) bind to taste receptors on the microvilli and produce a depolarizing receptor potential in the receptor cell.

III. Motor Systems

A. Motor unit

  —is a **single motoneuron and the muscle fibers that it innervates**. For **fine control** (e.g., eye movement), a single motoneuron innervates only a few muscle fibers. For **larger movements** (postural muscles), a single motoneuron may innervate thousands of muscle fibers.

  —The **motoneuron pool** is the group of motoneurons that innervates fibers within the same muscle.

—The force of muscle contraction is graded by **recruitment** of additional motor units (**size principle**).

1. **Small motoneurons**
   —innervate a few muscle fibers.
   —have the lowest thresholds and, therefore, **fire first**.
   —generate the **smallest force**.

2. **Large motoneurons**
   —innervate many muscle fibers.
   —have the highest thresholds and, therefore, **fire last**.
   —generate the **largest force**.

B. **Muscle sensors**

1. **Types of muscle sensors** (see Table 2-4)
   a. **Muscle spindles** (Ia) are arranged in parallel with extrafusal fibers and detect both **static and dynamic changes in muscle length**.
   b. **Golgi tendon organs** (Ib) are arranged in series with extrafusal muscle fibers and detect **muscle tension**.
   c. **Pacinian corpuscles** (II) are distributed throughout muscle and detect **vibration**.
   d. **Free nerve endings** (III, IV) detect **noxious stimuli**.

2. **Types of muscle fibers**
   a. **Extrafusal fibers**
      —make up the bulk of muscle.
      —are stimulated by α-motoneurons.
      —provide the **force for muscle contraction**.
   b. **Intrafusal fibers**
      —are smaller than extrafusal muscle fibers.
      —are encapsulated in sheaths to form **muscle spindles**.
      —run in parallel with extrafusal fibers, but do not run the entire length of the muscle.

3. **Muscle spindles**
   —are distributed throughout muscle.
   —consist of small, encapsulated intrafusal fibers connected in parallel with large (force-generating) extrafusal fibers.
   —The finer the movement required, the greater the number of muscle spindles in a muscle.
   a. **Types of intrafusal fibers in muscle spindles** (Figure 2-7)
      (1) **Nuclear bag fibers**
         —detect the rate of change in muscle length (fast, **dynamic** changes).
         —are innervated by group Ia afferents.
      (2) **Nuclear chain fibers**
         —detect **static** changes in muscle length.
         —are innervated by group II afferents.
   b. **How the muscle spindle works** (see Figure 2-7)
      —Muscle spindle reflexes oppose (correct for) increases in muscle length (stretch).

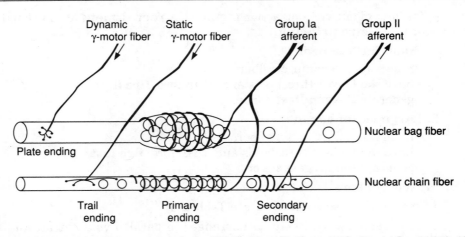

Figure 2-7. Organization of the muscle spindle. (Modified from Matthews PBC: *Physiol Rev* 44:219, 1964)

(1) Sensory information about muscle length is received by the group Ia (velocity) and group II (static) afferent fibers.

(2) When a muscle is lengthened (stretched), the muscle spindle is stretched, stimulating the group Ia and group II afferent fibers.

(3) Stimulation of the Ia afferents stimulates α-motoneurons in the spinal cord, which in turn cause contraction and shortening of the muscle. Thus, the original stretch is opposed and muscle length is maintained.

c. Function of the γ-motoneurons

—innervate intrafusal muscle fibers.

—adjust the sensitivity of the muscle spindle, so that it will **respond appropriately during muscle contraction.**

—**α-Motoneurons and γ-motoneurons are coactivated** so that muscle spindles remain sensitive to changes in muscle length during contraction.

C. Muscle reflexes (Table 2-7)

1. Stretch (myotatic) reflex—knee jerk (Figure 2-8)

—is **monosynaptic.**

a. Muscle is stretched, which stimulates **Ia** afferent fibers.

Table 2-7. Summary of Muscle Reflexes

Reflex	Number of synapses	Stimulus	Afferent fibers	Response
Stretch reflex (knee jerk)	Monosynaptic	Muscle is stretched	Ia	Contraction of the muscle
Golgi tendon reflex (clasp-knife)	Disynaptic	Muscle contracts	Ib	Relaxation of the muscle
Flexion–withdrawal reflex (after touching a hot stove)	Polysynaptic	Pain	II, III, and IV	Ipsilateral flexion; contralateral extension

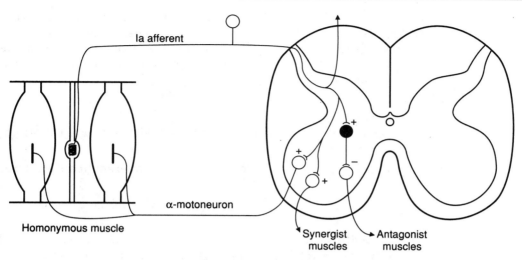

Figure 2-8. The stretch reflex.

b. Ia afferents synapse directly on α-motoneurons in the spinal cord. The pool of α-motoneurons that is activated causes contraction in the homonymous muscle.

c. Stimulation of α-motoneurons causes contraction in the muscle that was stretched. The muscle shortens, decreasing the stretch on the muscle spindle and returning it to its original length.

d. At the same time, synergistic muscles are activated and antagonistic muscles are inhibited.

e. Example of knee-jerk reflex. Tapping on the patellar tendon causes the quadriceps to stretch. Stretch of the quadriceps stimulates Ia afferent fibers that activate α-motoneurons to make the quadriceps contract, all of which forces the lower leg to extend. **Increases in γ-motoneuron activity** will increase the sensitivity of the muscle spindle and therefore exaggerate the knee-jerk reflex.

2. Golgi tendon reflex (inverse myotatic)

 –is **disynaptic**.

 –is the inverse of the stretch reflex.

a. Active muscle contraction stimulates the Golgi tendon organs and **Ib** afferent fibers.

b. The Ib afferents stimulate inhibitory interneurons in the spinal cord. These inhibitory interneurons inhibit α-motoneurons, causing relaxation of the muscle that was originally contracted.

c. At the same time, antagonistic muscles are excited.

d. Clasp-knife reflex, an exaggerated form of the Golgi tendon reflex, can occur with **disease of corticospinal tracts** (spasticity).

—**For example**, if the arm is hypertonic, the increased sensitivity of the muscle spindles in the extensor muscles (triceps) causes resistance to flexion of the arm. Eventually tension in the triceps increases to the point where it will activate the Golgi tendon reflex, the triceps relax, and the arm flexes like a jackknife.

3. Flexor withdrawal reflex

—is **polysynaptic**.

—Somatosensory and pain afferent fibers elicit withdrawal of the stimulated body part from the noxious stimulus.

a. Pain (e.g., touching a hot stove) stimulates the afferent fibers of groups II, III, and IV.

b. The afferent fibers synapse polysynaptically via interneurons onto motoneurons in the spinal cord.

c. On the ipsilateral side of the pain stimulus, flexors are stimulated and extensors are inhibited, and the arm is jerked away from the stove. **On the contralateral side**, flexors are inhibited and extensors are stimulated (**crossed extension reflex**) to maintain balance.

d. As a result of persistent neural activity in the polysynaptic circuits, there is an **afterdischarge**, which does not allow the muscle to relax for some time.

D. Spinal organization of motor systems

1. Convergence

—occurs when a single motoneuron receives its input from many muscle spindle Ia afferents in the homonymous muscle.

—produces **spatial summation** because a single input would not bring the muscle to threshold, but multiple inputs will.

—can also produce **temporal summation** when inputs arrive in rapid succession.

2. Divergence

—occurs when the muscle spindle Ia afferent fibers project to all of the motoneurons that innervate the homonymous muscle.

3. Recurrent inhibition (Renshaw cells)

—Renshaw cells are inhibitory cells in the ventral horn of the spinal cord.

—They receive input from collateral axons of motoneurons and, when stimulated, negatively feedback (inhibit) on the motoneuron.

E. Brainstem control of posture

1. Motor centers and pathways

—**Pyramidal tracts** (corticospinal and corticobulbar) pass through the medullary pyramids.

—All others are **extrapyramidal tracts** and originate primarily in the following structures of the brainstem:

a. Rubrospinal tract

—originates in the red nucleus and projects to interneurons in the lateral spinal cord.

—Stimulation of the red nucleus produces **stimulation of flexors** and **inhibition of extensors**.

b. Pontine reticulospinal tract

—originates in nuclei in the pons and projects to the ventromedial spinal cord.

—Stimulation has a general **stimulatory effect on both extensors and flexors**, with the **predominant effect on extensors**.

c. Medullary reticulospinal tract

—originates in the medullary reticular formation and projects to spinal cord interneurons in the intermediate gray area.

—Stimulation has a general **inhibitory effect on both extensors and flexors**, with the **predominant effect on extensors**.

d. Lateral vestibulospinal tract

—originates in Deiters' nucleus and projects to ipsilateral motoneurons and interneurons.

—Stimulation causes a powerful **stimulation of extensors** and **inhibition of flexors**.

e. Tectospinal tract

—originates in the superior colliculus and projects to the cervical spinal cord.

—is involved in the **control of neck muscles**.

2. Effects of transections of the spinal cord

a. Paraplegia

—is the loss of voluntary movements below the level of the lesion.

—results from interruption of descending pathways from motor centers in the brainstem and higher centers.

b. Loss of conscious sensation below the level of the lesion

c. Initial loss of reflexes—spinal shock

—Immediately following transection, there is loss of the excitatory influence from α- and γ-motoneurons; **limbs become flaccid** and **reflexes are absent**. With time, partial recovery and return of reflexes or even hyperreflexia will occur.

(1) If the **lesion is at C7**, there will be loss of sympathetic tone to the heart and, as a result, heart rate will slow and arterial pressure will decrease.

(2) If the **lesion is at C3**, breathing will stop because the respiratory muscles have been disconnected from control centers in the brainstem.

(3) If the **lesion is at C1** (e.g., as a result of hanging), there is certain death.

3. Effects of transections above the spinal cord

a. Lesions above the lateral vestibular nucleus

—cause **decerebrate rigidity** because of removal of inhibition from higher centers, resulting in excitation of α- and γ-motoneurons and rigid posture.

b. Lesions above the pontine reticular formation

–cause **decerebrate rigidity** because of removal of central inhibition from the pontine reticular formation, resulting in excitation of α- and γ-motoneurons and rigid posture.

c. Lesions above the red nucleus

–result in decorticate posturing and intact tonic neck reflexes.

F. Cerebellum—central control of movement

1. Functions of the cerebellum

a. Control of balance and eye movements (vestibulocerebellum)

b. Planning and initiation of movement (pontocerebellum)

c. Synergy—control of rate, force, range, and direction of movement (spinocerebellum)

2. Layers of the cerebellar cortex

a. Granular layer (innermost)

–contains granule cells, Golgi type II cells, and glomeruli.

–In the **glomeruli**, axons of mossy fibers form synaptic connections on dendrites of granular and Golgi II cells.

b. Purkinje cell layer (middle)

–contains Purkinje cells.

–**output is always inhibitory**.

c. Molecular layer (outermost)

–contains stellate and basket cells, dendrites of Purkinje and Golgi II cells, and parallel fibers (axons of granule cells).

3. Connections in the cerebellar cortex

a. Input to the cerebellar cortex

(1) Climbing fibers

–originate from a **single region** of the medulla (inferior olive).

–make multiple synapses onto Purkinje cells, resulting in high-frequency bursts or **complex spikes**.

–play a role in cerebellar **motor learning**.

(2) Mossy fibers

–originate from **many centers** in the brainstem and spinal cord.

–make multiple synapses on Purkinje fibers via interneurons.

–synapse on granule cells (glomerular complex). The axons of granule cells bifurcate and give rise to **parallel fibers**. The parallel fibers excite multiple Purkinje cells as well as inhibitory interneurons (basket, stellate, Golgi II).

b. Output of the cerebellar cortex

–**Purkinje cells** are the **only output from the cerebellar cortex**.

–Output of the Purkinje cells is **always inhibitory**; the neurotransmitter is **γ-aminobutyric acid (GABA)**.

–The output projects to deep cerebellar nuclei and to vestibular nucleus. This inhibitory output **modulates** the output of the cerebellum and regulates rate, range, and direction of movement (**synergy**).

c. Clinical disorders of the cerebellum— ataxia

–results in **lack of coordination**, including delay in initiation of movement, poor execution of a sequence of movements, and inability to perform rapid alternating movements (**dysdiadochokinesia**).

–**Intention tremor** occurs during attempts to perform voluntary movements.

–**Rebound phenomenon** is the inability to stop a movement.

G. Basal ganglia—control of movement

–consists of the **striatum, globus pallidus, subthalamic nuclei,** and **substantia nigra**.

–modulates thalamic outflow to the motor cortex to **plan and execute smooth movements**.

–Many of the synaptic connections are inhibitory and use **GABA** as their neurotransmitter.

–Lesions of basal ganglia include:

1. Lesions of the globus pallidus

–result in inability to maintain postural support.

2. Lesions of the subthalamic nucleus

–are caused by release of inhibition on the contralateral side.

–result in wild, flinging movements (e.g., hemiballismus).

3. Lesions of the striatum

–are caused by release of inhibition.

–result in quick, continuous, and uncontrollable movements.

–are seen in patients with Huntington's disease.

4. Lesions of the substantia nigra

–are caused by **destruction of dopaminergic neurons**.

–Because dopamine is an excitatory neurotransmitter, there is overactivity of inhibitory pathways from the striatum to the globus pallidus.

–Symptoms include **lead-pipe rigidity, tremor,** and **reduced voluntary movement**.

–are seen in patients with **Parkinson's disease.**

H. Motor cortex

1. Premotor cortex and supplementary motorcortex (area 6)

–are responsible for **generating a plan for movement**, which is then transferred to the primary motor cortex for execution.

–The supplementary motor cortex programs complex motor sequences and is active during **"mental rehearsal"** for a movement.

2. Primary motor cortex (area 4)

–is responsible for the **execution of movement**. Programmed patterns of motoneurons are activated in the motor cortex. Excitation of upper motoneurons in the motor cortex is transferred to the brainstem and spinal cord where the lower motoneurons are activated, resulting finally in voluntary movement.

–is somatotopically organized (**motor homunculus**). Epileptic events in the primary motor cortex cause **Jacksonian seizures**, which illustrate the somatotopic organization.

IV. Higher Functions of the Cerebral Cortex

A. Electroencephalographic (EEG) findings

–EEG waves consist of alternating excitatory and inhibitory synaptic potentials in the pyramidal cells of the cerebral cortex.

–A **cortical evoked potential** is an EEG change and reflects synaptic potentials evoked in large numbers of neurons.

–In alert individuals, **beta waves** predominate. Upon relaxation, **alpha waves** predominate. During sleep, **slow waves** predominate, muscles relax, and heart rate and blood pressure decrease.

B. Sleep

1. Sleep–wake cycles occur in a **circadian rhythm** with a period of about 24 hours. The circadian periodicity is thought to be driven by the suprachiasmatic nucleus of the **hypothalamus**, which receives input from the retina.

2. **Rapid eye movement (REM) sleep** occurs every 90 minutes. During REM sleep, the EEG resembles that of a person who is awake. Most **dreams** occur during REM sleep. Use of **benzodiazepines** and **increasing age** decrease the duration of REM sleep.

C. Language

–Information is transferred between the two hemispheres of the cerebral cortex through the corpus callosum.

–The **right hemisphere** is more dominant than the left hemisphere in facial expression, intonation, body language, and spatial tasks.

–The **left hemisphere** is usually dominant with respect to **language**, (even in left-handed people). Lesions of the left hemisphere cause **aphasia**.

1. Damage to **Wernicke's area** causes **sensory aphasia**, in which there is difficulty understanding written or spoken language.

2. Damage to **Broca's area** causes **motor aphasia**, in which speech and writing are affected but understanding is intact.

D. Learning and memory

–**Short-term memory** involves synaptic changes.

–**Long-term memory** involves structural changes in the nervous system and is more resistant.

–Bilateral lesions of the **hippocampus** block the ability to form new long-term memories.

V. Blood–Brain Barrier and Cerebrospinal Fluid

A. Anatomy of the blood–brain barrier

–It is the barrier between cerebral capillary blood and cerebrospinal fluid (CSF). CSF fills the ventricles and the subarachnoid space.

–consists of the **endothelial cells of the cerebral capillaries** and the **choroid plexus epithelium**.

B. Formation of CSF by the choroid plexus epithelium

 —Lipid-soluble substances (CO_2 and O_2) and H_2O freely cross the blood–brain barrier and equilibrate between blood and CSF.
 —Other substances are transported by carriers in the choroid plexus epithelium. They may be secreted from blood into CSF or absorbed from CSF into blood.
 —**Protein** and **cholesterol** are excluded from the CSF because of their large molecular size.
 —The composition of CSF is approximately the same as the interstitial fluid of the brain, but differs significantly from blood (Table 2-8).

C. Functions of the blood–brain barrier

1. It **maintains a constant environment for neurons in the CNS** and protects the brain from endogenous or exogenous toxins.

2. It **prevents escape of neurotransmitters** from their functional sites in the CNS into the general circulation.

3. **Drugs** penetrate the blood–brain barrier to varying degrees. **For example**, non-ionized (lipid-soluble) drugs will cross more readily than ionized (non-lipid-soluble) drugs.

   —**Inflammation, irradiation,** and **tumors** may destroy the blood–brain barrier and permit entry into the brain of substances that usually are excluded (e.g., antibiotics or radiolabelled markers).

VI. Temperature Regulation

A. Sources of heat gain and heat loss from the body

1. Heat-generating mechanisms—response to cold

a. **Thyroid hormone** increases metabolic rate and heat production by stimulating the Na^+–K^+ adenosine triphosphatase (ATPase).

b. **Cold temperatures activate the sympathetic nervous system** and, via activation of β receptors in **brown fat,** increase metabolic rate and heat production.

c. **Shivering** is the most potent mechanism for increasing heat production. Cold temperatures activate the shivering response orchestrated by the posterior hypothalamus; α- and γ-motoneurons are activated, causing contraction of skeletal muscle and heat production.

Table 2-8. Comparison of Cerebrospinal Fluid and Blood Concentrations

[CSF] ≈ [Blood]	[CSF] < [Blood]	[CSF] > [Blood]
Na^+	K^+	Mg^{2+}
Cl^-	Ca^{2+}	Creatinine
HCO_3^-	Glucose	
Osmolarity	Cholesterol**	
	Protein**	

*CSF = cerebrospinal fluid.
**Negligible concentration in CSF.

2. Heat-loss mechanisms—response to heat

   a. Heat loss by **radiation** and **convection** increases when the ambient temperature increases.

   —The response is orchestrated by the posterior hypothalamus.

   —Increases in temperature cause a **decrease in sympathetic tone to skin blood vessels**, increasing blood flow through arterioles and increasing **arteriovenous shunting of blood to venous plexus** near the surface of the skin. Shunting of warm blood to the surface of the skin increases heat loss by radiation and convection.

   b. Heat loss by **evaporation** depends upon the activity of **sweat glands**, which are under **sympathetic muscarinic** control.

B. Hypothalamic set point for body temperature

   1. Temperature sensors on the skin and in the hypothalamus (which "read" the core temperature) relay information to the **anterior hypothalamus**.

   2. The anterior hypothalamus compares the core temperature to the **set-point temperature**.

   a. If the core temperature is below the set point, heat-generating mechanisms (increased metabolism, shivering, vasoconstriction of skin blood vessels) are activated by the posterior hypothalamus.

   b. If the **core temperature is above the set point**, mechanisms for heat loss (vasodilation of the skin blood vessels, increased sympathetic outflow to the sweat glands) are activated by the posterior hypothalamus.

   3. Cold and pyrogens **increase the set-point temperature**. Core temperature will be seen as lower than the new set-point temperature by the anterior hypothalamus. As a result, heat-generating mechanisms (e.g., shivering) will be initiated.

C. Fever

   1. Pyrogens increase the production of **interleukin-1** (IL-1) in phagocytic cells. IL-1 acts on the anterior hypothalamus to increase the production of prostaglandins, which **increase the set-point temperature,** setting in motion the heat-generating mechanisms that raise body temperature and produce **fever**.

   2. Aspirin reduces fever by **inhibiting cyclooxygenase**, thereby inhibiting the production of prostaglandins; therefore, aspirin **decreases the set-point temperature**. In response, mechanisms are activated that cause heat loss (sweating and vasodilation).

   3. Steroids reduce fever by blocking the release of arachidonic acid from brain phospholipids, thereby preventing the production of prostaglandins.

D. Heat exhaustion and heat stroke

   1. Heat exhaustion is caused by excessive sweating, which causes decreased blood volume and, as a result, decreased arterial blood pressure and syncope.

   2. Heat stroke occurs when body temperature increases to the point of tissue damage. The normal response to increased ambient temperature (sweating) is impaired, and core temperature will rise further.

E. Hypothermia

   —results when the ambient temperature is so low that heat-generating mechanisms (e.g., shivering, metabolism) cannot adequately maintain core temperature near the set point.

F. Malignant hyperthermia

   —is caused in susceptible individuals by inhalation anesthetics.

   —is characterized by a massive increase in oxygen consumption and heat production by skeletal muscle, which causes a rapid rise in body temperature.

Review Test

Directions: Each of the numbered items or incomplete statements in this section is followed by answers or by completions of the statement. Select the **one** lettered answer or completion that is **best** in each case.

1. The inability to perform rapidly alternating movements (dysdiadochokinesia) is associated with lesions of the

(A) premotor cortex
(B) motor cortex
(C) cerebellum
(D) substantia nigra
(E) medulla

2. Which of the following responses is mediated by parasympathetic muscarinic receptors?

(A) Dilation of bronchiolar smooth muscle
(B) Erection
(C) Ejaculation
(D) Constriction of gastrointestinal sphincters
(E) Increased cardiac contractility

3. Which of the following is a property of C fibers?

(A) Have the slowest conduction velocity of any nerve fiber type
(B) Have the largest diameter of any nerve fiber type
(C) Are afferent nerves from muscle spindles
(D) Are afferent nerves from Golgi tendon organs
(E) Are preganglionic autonomic fibers

4. When compared with the cones of the retina, the rods

(A) are more sensitive to low-intensity light
(B) adapt before the cones in the dark
(C) are in highest concentration on the fovea
(D) are primarily involved in color vision

5. Which of the following statements best describes the basilar membrane of the organ of Corti?

(A) The apex responds better to low frequencies than the does the base
(B) The base is wider than the apex
(C) The base is more compliant than the apex
(D) High frequencies produce maximal displacement of the basilar membrane near the helicotrema
(E) The apex is relatively stiff compared to the base

6. A lesion of the chorda tympani nerve would most likely result in

(A) impaired olfactory function
(B) impaired vestibular function
(C) impaired auditory function
(D) impaired taste function
(E) nerve deafness

7. Complete transection of the spinal cord at the level of T1 would most likely result in

(A) temporary loss of stretch reflexes below the lesion
(B) temporary loss of conscious proprioception below the lesion
(C) permanent loss of voluntary control of movement above the lesion
(D) permanent loss of consciousness above the lesion

8. All of the following are steps in photoreception in the rods EXCEPT

(A) light converts 11-*cis* rhodopsin to all-*trans* rhodopsin
(B) *meta*-rhodopsin II activates transducin
(C) cyclic GMP levels decrease
(D) rods depolarize
(E) decreased release of neurotransmitter

9. Pathogens that produce fever cause all of the following EXCEPT

(A) increased production of interleukin-1 (IL-1)
(B) increased set-point temperature in the hypothalamus
(C) shivering
(D) vasodilation of blood vessels in the skin

10. Which of the following is a characteristic of nuclear bag fibers?

(A) They are one type of extrafusal muscle fiber
(B) They detect dynamic changes in muscle length
(C) They give rise to group Ib afferents
(D) They are innervated by α-motoneurons

11. Which type of cell in the visual cortex responds best to a moving bar of light?

(A) Simple
(B) Complex
(C) Hypercomplex
(D) Bipolar
(E) Ganglion

12. Which adrenergic receptor produces its stimulatory effects by formation of IP_3 and an increase in intracellular $[Ca^{2+}]$?

(A) α_1 Receptors
(B) β_1 Receptors
(C) β_2 Receptors
(D) Muscarinic receptors
(E) Nicotinic receptors

13. The excessive muscle tone produced in decerebrate rigidity can be reversed by

(A) stimulation of Ia afferents
(B) cutting the dorsal roots
(C) transection of cerebellar connections to the lateral vestibular nucleus
(D) stimulation of α-motoneurons
(E) stimulation of γ-motoneurons

14. Which of the following parts of the body has cortical motoneurons with the largest representation on primary motor cortex (area 4)?

(A) Shoulder
(B) Ankle
(C) Fingers
(D) Elbow
(E) Knee

15. Which of the following would produce maximum excitation of hair cells in the right horizontal semicircular canal?

(A) Hyperpolarization of the hair cells
(B) Bending the stereocilia away from the kinocilia
(C) Rapid ascent in an elevator
(D) Rotating the head to the right

16. All of the following statements about the olfactory system are true EXCEPT

(A) the receptor cells are neurons
(B) the receptor cells are continuously replaced
(C) axons of CN I are A-delta fibers
(D) axons from receptor cells synapse directly onto cells in the brain
(E) fractures of the cribriform plate can cause anosmia

17. Sensory receptor potentials

(A) are action potentials
(B) always bring the membrane potential of a receptor cell toward threshold
(C) always bring the membrane potential of a receptor cell away from threshold
(D) are graded in size, depending on stimulus intensity
(E) are all-or-none

18. A ballet dancer spins to the left. During the spin, her eyes snap quickly to the left. This fast eye movement is

(A) nystagmus
(A) post-rotatory nystagmus
(C) ataxia
(D) aphasia

19. Which of the following has a much lower concentration in cerebrospinal fluid (CSF) than in cerebral capillary blood?

(A) Na^+
(B) K^+
(C) Osmolarity
(D) Protein
(E) Mg^{2+}

Directions: The group of items in this section consists of lettered options followed by a set of numbered items. For each item, select the **one** lettered option that is most closely associated with it. Each lettered option may be selected once, more than once, or not at all.

Questions 20–25

Match each numbered phenomenon below to the autonomic receptors that mediate it.

(A) Adrenergic α receptors
(B) Adrenergic β_1 receptors
(C) Adrenergic β_2 receptors
(D) Cholinergic muscarinic receptors
(E) Cholinergic nicotinic receptors

20. Increased heart rate

21. Secretion of epinephrine by the adrenal medulla

22. Ion channels for Na^+ and K^+ at the neuromuscular junction

23. Blocked by hexamethonium at ganglia but not at the neuromuscular junction

24. Atropine decreases gastrointestinal motility by blocking this receptor

25. Low concentrations of epinephrine released from the adrenal medulla cause vasodilation

Questions 26–28

Match each numbered deficit to the appropriate optic pathway.

(A) Optic nerve
(B) Optic chiasm
(C) Optic tract
(D) Geniculocalcarine tract

26. Cutting on the left side causes total blindness in the left eye

27. Cutting on the right side causes blindness in the temporal field of the left eye and the nasal field of the right eye

28. Cutting causes blindness in the temporal fields of the left and right eyes

Questions 29 and 30

Match each numbered description below with the correct structure.

(A) Motor cortex
(B) Premotor and supplementary motor cortex
(C) Prefrontal cortex
(D) Basal ganglia
(E) Cerebellum

29. Active during programming and "mental rehearsal" of complex motor sequences in the absence of movement

30. Primary function is to coordinate rate, range, force, and direction of movement

Questions 31 and 32

Match each numbered description with the correct reflex.

(A) Stretch reflex (myotatic)
(B) Golgi tendon reflex (inverse myotatic)
(C) Flexion-withdrawal reflex
(D) Subliminal occlusion reflex

31. Polysynaptic excitation of contralateral extensors

32. Monosynaptic excitation of ipsilateral homonymous muscle

Questions 33–35

Match each numbered description below to the correct nerve fiber.

(A) α-Motoneurons
(B) γ-Motoneurons
(C) Group Ia fibers
(D) Group Ib fibers

33. Detect increases in muscle tension

34. Stimulation leads to contraction of the bulk of skeletal muscle

35. Muscle stretch leads to a direct increase in its firing rate

Answers and Explanations

1–C. Coordination of movement (synergy) is the function of the cerebellum. Lesions of the cerebellum cause ataxia, lack of coordination, poor execution of movement, delay in initiation of movement, and inability to perform rapidly alternating movements. The premotor and motor cortices plan and execute movements. Lesions of the substantia nigra, a component of the basal ganglia, result in tremors, lead-pipe rigidity, and poor muscle tone (Parkinson's disease).

2–B. Erection is a parasympathetic muscarinic response. Dilation of bronchioles, ejaculation, constriction of gastrointestinal sphincters, and increased cardiac contractility are all sympathetic α or β responses.

3–A. C fibers (slow pain) are the smallest nerve fibers and therefore have the slowest conduction velocity.

4–A. Of the two types of photoreceptors, the rods are more sensitive to low-intensity light and therefore are more important than the cones for night vision. They adapt to darkness after the cones. Rods are not present in the fovea. The cones are primarily involved in color vision.

5–A. Sound frequencies can be encoded by the organ of Corti because of differences in properties along the basilar membrane. The base of the basilar membrane is narrow and stiff, and hair cells on it are activated by high frequencies. The apex of the basilar membrane is wide and compliant, and hair cells on it are activated by low frequencies.

6–D. The chorda tympani (CN VII) is involved in taste; it innervates the anterior two-thirds of the tongue.

7–A. Transection of the spinal cord causes "spinal shock" and loss of all reflexes below the level of the lesion. These reflexes, which are local circuits within the spinal cord, will return with time or become hypersensitive. Proprioception is permanently (rather than temporarily) lost because of the interruption of sensory nerve fibers. Fibers above the lesion are intact.

8–D. Photoreception involves the following steps. Light converts 11-*cis* rhodopsin to all-*trans* rhodopsin, which is converted to such intermediates as *meta*-rhodopsin II. *Meta*-rhodopsin II activates a stimulatory G protein (transducin), which activates a phosphodiesterase. Phosphodiesterase breaks down cyclic GMP, so intracellular cyclic GMP levels decrease, causing closure of Na^+ channels in the rod cell membrane and hyperpolarization. Hyperpolarization of the rod cell membrane inhibits the release of neurotransmitter. If the neurotransmitter is excitatory, then the bipolar cell will be hyperpolarized (inhibited). If the neurotransmitter is inhibitory, then the bipolar cell will be depolarized (excited).

9–D. Pathogens release IL-1 from phagocytic cells. IL-1 then acts to increase production of prostaglandins, ultimately raising the temperature set point in the anterior hypothalamus. The hypothalamus now "thinks" that body temperature is too low (since core temperature is lower than the new set-point temperature) and initiates mechanisms for generating heat—shivering, vasoconstriction, and shunting of blood away from venous plexus near the skin surface.

10–B. Nuclear bag fibers are one type of intrafusal muscle fiber that make up muscle spindles. They detect dynamic changes in muscle length, give rise to group Ia afferent fibers, and are innervated by γ-motoneurons. The other type of intrafusal fiber, the nuclear chain fiber, detects static changes in muscle length.

11–B. Complex cells respond to moving bars or edges with the correct orientation. Simple cells respond to stationary bars, and hypercomplex cells respond to lines, curves, and angles. Bipolar and ganglion cells are in the retina, not in the visual cortex.

12–A. Adrenergic α_1 receptors produce physiologic actions by stimulating formation of IP_3 and causing a subsequent increase in intracellular $[Ca^{2+}]$. Both types of β receptors act by stimulating adenylate cyclase and production of cyclic AMP. Muscarinic and nicotinic receptors are cholinergic.

13–B. Decerebrate rigidity is caused by increased reflex muscle spindle activity. Stimulation of Ia afferents would enhance, not diminish, this reflex activity. Cutting the dorsal roots would block the reflexes. Stimulation of α- and γ-motoneurons would stimulate muscles directly.

14–C. Representation on the motor homunculus is greatest for those structures involved in the most complicated movements—the fingers, hands, and face.

15–D. The semicircular canals are involved in angular acceleration or rotation. Hair cells of the right semicircular canal are excited (depolarized) when there is rotation to the right. This rotation causes bending of the stereocilia toward the kinocilia, which produces depolarization of the hair cell. Ascent in an elevator would activate the saccules, which detect linear acceleration.

16–C. CN I innervates the olfactory epithelium, and its axons are C fibers. Fracture of the cribriform plate can tear the delicate olfactory nerves and eliminate the sense of smell (anosmia). Olfactory receptor cells are unique in that they are true neurons, which are continuously replaced from undifferentiated stem cells.

17–D. Receptor potentials are graded potentials that may bring the membrane potential of the receptor cell either toward (depolarizing) or away from (hyperpolarizing) threshold. They are not action potentials, but action potentials may result if the membrane potential reaches threshold.

18–A. The fast eye movement during a spin is nystagmus and is in the same direction as the rotation. After the spin, post-rotatory nystagmus occurs in the opposite direction from the spin.

19–D. CSF is similar in composition to the interstitial fluid of the brain. Therefore, it is similar to an ultrafiltrate of plasma and has a very low protein concentration since large protein molecules cannot cross the blood–brain barrier. There are other differences in composition between CSF and blood that are created by transporters in the choroid plexus, but the low protein concentration of CSF is the most dramatic.

20–B. Heart rate is increased by the stimulatory effect of norepinephrine on β_1 receptors in the SA node. There are also sympathetic β_1 receptors in the heart that regulate contractility.

21–E. Preganglionic parasympathetic fibers synapse on the chromaffin cells of the adrenal medulla at a nicotinic receptor, releasing epinephrine (and to a lesser extent, norepinephrine) into the circulation.

22–E. Nicotinic receptors at the neuromuscular junction are not only the postsynaptic receptors for ACh, but are also the ion channels for Na^+ and K^+; when they are opened by ACh, they depolarize the muscle end plate (the end plate potential).

23–E. Hexamethonium is a nicotinic blocker, but acts only at ganglionic (not neuromuscular junction) nicotinic receptors. This pharmacologic distinction emphasizes that nicotinic receptors at these two locations, while similar, are not identical.

24–D. Atropine is a specific inhibitor of muscarinic receptors for ACh. Atropine would be expected to block all effects of ACh that are mediated by muscarinic receptors. Therefore, atropine will decrease gastrointestinal motility, decrease bladder tone, dilate bronchiolar smooth muscle, and increase heart rate.

25–C. β_2 Receptors on vascular smooth muscle produce vasodilation. α Receptors on vascular smooth muscle produce vasoconstriction. Because β_2 receptors are more sensitive to epinephrine than are α receptors, low doses of epinephrine produce vasodilation, and high doses produce vasoconstriction.

26–A. Cutting the optic nerves from the left eye causes blindness in the left eye because the fibers have not yet crossed at the optic chiasm.

27–C. Fibers from the left temporal field and the right nasal field ascend together in the right optic tract.

28–B. Optic nerve fibers from both temporal receptor fields cross at the optic chiasm.

29–B. The premotor cortex (area 6) is responsible for generating a plan for movement before movement occurs. The supplementary motor cortex is responsible for the "mental rehearsal" of the movement.

30–E. Output of Purkinje cells of the cerebellar cortex to deep cerebellar nuclei is inhibitory. This modulates movement and is responsible for the coordination that allows one to "catch a fly."

31–C. Flexion-withdrawal is a polysynaptic reflex that is used when a person touches a hot stove or steps on a tack. On the ipsilateral side, there is flexion (withdrawal); on the contralateral side, there is extension to maintain balance.

32–A. The stretch reflex is the monosynaptic response to stretching of a muscle. The reflex produces contraction and shortening of the muscle that was originally stretched (homonymous muscle).

33–D. Group Ib fibers are the afferent fibers from Golgi tendon organs that detect increases in muscle tension. The Golgi tendon reflex (inverse myotatic) compensates for the increased tension by causing relaxation of the muscle.

34–A. Extrafusal fibers constitute the bulk of muscle and are innervated by α-motoneurons.

35–C. Ia afferent fibers innervate intrafusal fibers of the muscle spindle. When the intrafusal fibers are stretched, the Ia fibers fire and activate the stretch reflex, which causes the muscle to shorten back to its resting length.

3

Cardiovascular Physiology

I. Hemodynamics

A. Components of the vasculature

1. Systemic arteries

—deliver oxygenated blood to the organs.

—are thick-walled and under high pressure.

—The blood volume contained in systemic arteries is called the **stressed volume**.

2. Arterioles

—are the smallest branches of the arteries.

—are the **site of highest resistance in the cardiovascular system**.

—have a smooth muscle wall that is extensively innervated by autonomic fibers.

—Arteriolar resistance is regulated by the autonomic nervous system.

3. Capillaries

—have the **largest total cross-sectional and surface area**.

—are thin-walled.

—are the site of exchange of nutrients, water, and gases.

4. Venules

—are innervated by autonomic fibers.

5. Veins

—progressively merge to form larger veins, eventually the venae cavae (inferior and superior), to return blood to the heart.

—have the lowest pressure.

—contain the **highest proportion of the blood in the cardiovascular system**.

—The blood volume contained in the veins is called the **unstressed volume**.

B. Velocity of blood flow

—can be expressed by the following equation:

$$v = Q/A$$

v	velocity	cm/sec
Q	blood flow	ml/min
A	cross-sectional area	cm^2

—Velocity is directly proportional to blood flow and inversely proportional to the cross-sectional area at any level of the cardiovascular system.
—**For example, blood velocity** is higher in the aorta (small cross-sectional area) than in the sum of all the capillaries (large cross-sectional area). The lower velocity of blood flow in the capillaries optimizes the exchange of substances across the capillary wall.

C. Blood flow

—can be expressed by the following equation:

$$Q = \Delta P / R$$

or

$$\text{Cardiac output} = \frac{\text{mean arterial pressure—right atrial pressure}}{\text{total peripheral resistance (TPR)}}$$

Q	flow	cardiac output	ml/min
ΔP	pressure gradient	pressure gradient	mm Hg
R	resistance	total peripheral resistance	mm Hg/ml/min

—The equation for blood flow (or cardiac output) is analogous to Ohm's law for electrical circuits (I = V/R), where flow is analogous to current and pressure is analogous to voltage.
—The pressure gradient (ΔP) drives blood flow.
—Thus, blood flows from high pressure to low pressure, and flow is inversely proportional to the resistance of the blood vessels.

D. Resistance

—The Poiseuille equation gives factors that change the resistance of blood vessels.

$$R = \frac{8\eta l}{\pi r^4}$$

R	resistance
η	viscosity of blood
l	length of blood vessel
r^4	radius of blood vessel to fourth power

—Resistance is directly proportional to the viscosity of the blood. **For example,** increasing viscosity by increasing hematocrit will increase resistance and decrease flow.
—Resistance is directly proportional to length of vessel.

−Resistance is inversely proportional to the fourth power of the vessel radius. This is a powerful relationship. **For example,** if blood vessel radius decreases by a factor of 2, then resistance will increase by a factor of 16 (2^4), and flow will therefore decrease by a factor of 16.

1. Resistances in parallel or series

    a. Parallel resistance is illustrated by the systemic circulation; each organ is supplied by an artery that branches off the aorta. The total resistance of this parallel arrangement is expressed in the following equation:

$$\frac{1}{R_{total}} = \frac{1}{R_a} + \frac{1}{R_b} + \; .... \; \frac{1}{R_n}$$

    R_a, R_b, and R_n are the resistances of the renal, hepatic, and .... arteries, respectively. The total resistance is less than the resistance of any of the individual arteries.

    b. Series resistance is illustrated by the arrangement of blood vessels within a given organ. Each organ is supplied by a large artery, smaller arteries, then arterioles, capillaries, and veins arranged in series. The total resistance is the sum of the individual resistances, as expressed in the following equation:

$$R_{total} = R_{artery} + R_{arterioles} + R_{capillaries}$$

2. Laminar flow versus turbulent flow

    −Laminar flow is streamlined (in a straight line), turbulent flow is not.

    −**Reynold's number** predicts whether blood flow will be laminar or turbulent.

    −When Reynold's number is increased, there is a greater tendency for **turbulence,** which causes audible vibrations called **bruits**. Reynold's number (and therefore **turbulence) is increased** by the following factors:

    a. ↓ blood viscosity (e.g., ↓ hematocrit, **anemia**)

    b. ↑ blood velocity (e.g., **narrowing of a vessel**)

E. Capacitance (compliance)

−describes the **distensibility of blood vessels**.

−can be expressed by the following equation:

$$C = V/P$$

C	capacitance	ml/mm Hg
V	volume	ml
P	pressure	mm Hg

−is directly proportional to volume and inversely proportional to pressure.

−describes how volume changes in response to a change in pressure.

−is much greater for veins than arteries. As a result, more blood volume is contained in the veins (**unstressed volume**) than in the arteries (**stressed volume**).

−Changes in the capacitance of the veins produce changes in unstressed volume. For example, a decrease in venous capacitance decreases unstressed volume and increases stressed volume by shifting blood from the veins to the arteries.

–**Capacitance of the arteries decreases as a person ages;** therefore, the arteries become less distensible.

F. Pressure profile in blood vessels

–As blood flows through the systemic circulation, pressure falls progressively because of the resistance to blood flow.

–Thus, pressure is highest in the aorta and lowest in the venae cavae.

–The **largest fall in pressure occurs across the arterioles** because they are the site of highest resistance.

G. Arterial pressure

–is pulsatile.

–is not constant during a cardiac cycle.

1. Systolic pressure

–is the highest arterial pressure during a cardiac cycle.

–occurs when the heart contracts and blood is ejected into the arterial system.

2. Diastolic pressure

–is the lowest arterial pressure during a cardiac cycle.

–occurs when the heart is relaxed and blood is being returned to the heart via the veins.

3. Pulse pressure

–is the difference between systolic and diastolic pressures.

–The **most important determinant of pulse pressure is stroke volume**. As blood is ejected from the left ventricle into the arterial system, systolic pressure increases dramatically because of the relatively low capacitance of the arteries. Since diastolic pressure remains unchanged during ventricular systole, the pulse pressure increases to the same extent as does systolic pressure.

–**Decreases in capacitance, such as those that occur with the aging process, cause the pulse pressure to increase.**

4. Mean arterial pressure

–is the average arterial pressure with respect to time.

–can be calculated approximately as diastolic pressure plus one-third of pulse pressure.

II. Cardiac Electrophysiology

A. Electrocardiogram (Figure 3-1)

1. P wave

–represents depolarization of atrial muscle.

–does not include atrial repolarization, which is "buried" in the QRS complex.

2. PR interval

–is the interval from first atrial depolarization to the beginning of the Q wave (initial depolarization of the ventricle).

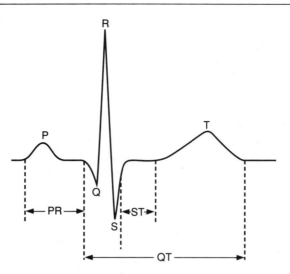

Figure 3-1. Normal electrocardiogram measured from lead II.

—increases if conduction velocity through the atrioventricular (AV) node is slowed (as in **heart block**).

—varies with heart rate: When heart rate increases, the PR interval decreases.

3. QRS complex

—represents depolarization of the ventricle.

4. QT interval

—is the interval from the beginning of the Q wave to the end of the T wave.

—represents the entire period of depolarization and repolarization of the ventricle.

5. ST segment

—is the segment from the end of the S wave to the beginning of the T wave.

—is isoelectric.

—represents the period when the entire ventricle is depolarized.

6. T wave

—is ventricular repolarization.

B. Cardiac action potentials

—The **resting membrane potential** is determined by the conductance to K^+ and approaches the K^+ equilibrium potential.

—**Inward current** brings positive charge into the cell and **depolarizes the membrane potential**.

—**Outward current** takes positive charge out of the cell and **hyperpolarizes the membrane potential**.

—The **role of the Na^+–K^+ adenosine triphosphatase (ATPase)** is to maintain ion gradients across the cell membrane.

1. Ventricle, atria, and Purkinje system (Figure 3-2)

–have stable resting membrane potentials of about -90 mV, which approaches the K$^+$ equilibrium potential.

–Action potentials are of long duration, especially in the ventricle, with a duration of 300 msec.

a. Phase 0

–is the **upstroke** of the action potential.

–is caused by a transient increase in Na$^+$ conductance. This increase results in an inward Na$^+$ current that depolarizes the membrane.

–At the peak of the action potential, the membrane potential approaches the equilibrium potential for Na$^+$.

b. Phase 1

–is a brief period of initial repolarization.

–**Initial repolarization** is caused by an outward current, in part because of K$^+$ ions moving out of the cell (favored by both chemical and electrical gradients) and in part because of a decrease in Na$^+$ conductance.

c. Phase 2

–is the **plateau** of the action potential.

–is caused by a **transient increase in Ca^{2+} conductance,** which results in an **inward Ca^{2+} current** and an increase in K$^+$ conductance. During the plateau, outward and inward currents are approximately equal, so the membrane potential is stable at the plateau level.

d. Phase 3

–is **repolarization**.

–During phase 3, Ca^{2+} conductance decreases, but K$^+$ conductance increases and therefore predominates.

–The high K$^+$ conductance results in a large outward K$^+$ current, which hyperpolarizes the membrane back toward the K$^+$ equilibrium potential.

e. Phase 4

–is the **resting membrane potential**.

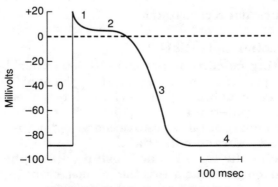

Figure 3-2. Ventricular action potential.

−is a period during which inward and outward currents are equal and the membrane potential approaches the K^+ equilibrium potential.

2. Sinoatrial (SA) node [Figure 3-3]

−is normally the **pacemaker** of the heart.
−**does not have a constant resting potential.**
−exhibits phase 4 depolarization or **automaticity**.
−The AV node and His-Purkinje systems are **latent pacemakers** that may exhibit automaticity and override the SA node if the SA node is not functioning. The intrinsic rate of phase 4 depolarization (and heart rate) is slower in the AV node and in the Purkinje system than in the SA node:

$$SA \text{ node} > AV \text{ node} > His\text{-}Purkinje$$

a. Phase 0

−is the **upstroke of the action potential**.
−is caused by an increase in Ca^{2+} conductance. This increase causes an **inward Ca^{2+} current** that drives the membrane potential toward the Ca^{2+} equilibrium potential.
−The **ionic basis for phase 0 is different from that in the ventricle** (where it is the result of an inward Na^+ current).

b. Phase 3

−is **repolarization**.
−is caused by an increase in K^+ conductance. This increase results in an outward K^+ current that causes repolarization of the membrane potential.

c. Phase 4

−is the **slow depolarization**.
−accounts for the **pacemaker activity** of the SA node (automaticity).
−is **caused by an increase in Na^+ conductance, which results in an inward Na^+ current called I_f.**
−I_f **is turned on by repolarization** of the membrane during the preceding action potential.

d. Phases 1 and 2

−are not present in the SA node action potential.

C. Conduction velocity

−reflects the time required for excitation to spread in heart tissue.

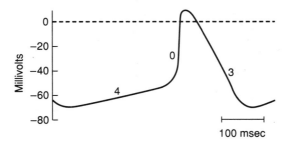

Figure 3-3. Sinoatrial (SA) nodal action potential.

–depends upon the **size of the inward current during the upstroke** of the action potential. The greater the inward current, the higher the conduction velocity.

–is fastest in the Purkinje system.

–is **slowest in the AV node (long PR interval on the ECG), allowing for filling of the ventricle prior to ventricular contraction**. If conduction velocity through the AV node is increased, then ventricular filling may be compromised.

D. Excitability (Figure 3-4)

–is the ability of the heart to initiate an action potential in response to an inward, depolarizing current.

–changes during the course of an action potential; these changes are described by **refractory periods**.

–reflects the recovery of the channels that carry the inward currents for the upstroke of the action potential.

1. Absolute refractory period (ARP)

–begins with the upstroke of the action potential and ends after the plateau.

–reflects the time during which **no action potential can be initiated**.

2. Effective refractory period (ERP)

–is slightly longer than the absolute refractory period.

–is the period during which a **conducted action potential cannot be elicited**.

3. Relative refractory period (RRP)

–is the period just following the absolute refractory period when repolarization is almost complete.

–is the period during which an **action potential can be elicited, but more than the usual inward current is required**.

E. Autonomic effects on heart rate and conduction velocity (Table 3-1)

–See III C for discussion of inotropic effects.

1. Definitions of chronotropic and dromotropic effects

a. Chronotropic effects

–are on heart rate.

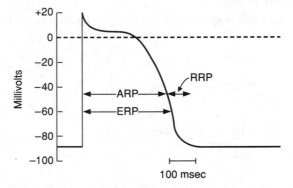

Figure 3-4. Absolute, effective, and relative refractory periods in the ventricle.

Table 3-1. Autonomic Effects on the Heart and Blood Vessels

	Sympathetic		Parasympathetic	
	Effect	**Receptor**	**Effect**	**Receptor**
Heart rate	↑	β_1	↓	Muscarinic
Conduction velocity (AV node)	↑	β_1	↓	Muscarinic
Contractility	↑	β_1	↓ (atria only)	Muscarinic
Vascular smooth muscle				
Skin, splanchnic	Constriction	α		
Skeletal muscle	Constriction	α		
	Relaxation	β_2		

–A **negative chronotropic effect** decreases heart rate by decreasing the rate of firing of the SA node.

–A **positive chronotropic effect** increases heart rate by increasing the rate of firing of the SA node.

b. Dromotropic effects

–are on conduction velocity, primarily in the AV node.

–A **negative dromotropic effect** decreases conduction velocity through the AV node, slowing the conduction of action potentials from atria to ventricles.

–A **positive dromotropic effect** increases conduction velocity through the AV node, speeding the conduction of action potentials from atria to ventricles.

2. Parasympathetic effects on heart rate and conduction velocity (vagus nerve)

–The SA node, atria, and AV node have parasympathetic innervation, but ventricles do not. The transmitter is **acetylcholine** (ACh) at a **muscarinic receptor**.

a. Negative chronotropic effect

–**decreases heart rate** by decreasing the rate of phase 4 depolarization.

–Fewer action potentials occur per unit time because threshold potential is reached more slowly.

–The **mechanism is a decrease in I_f,** the inward Na^+ current that is responsible for phase 4 depolarization in the SA node.

b. Negative dromotropic effect

–**decreases conduction velocity through the AV node.**

–**increases the PR interval.**

–Action potentials are conducted more slowly from atria to ventricles.

3. Sympathetic effects on heart rate and conduction velocity

–**Norepinephrine** is the **neurotransmitter at a β_1 receptor.**

–Innervation is throughout the heart.

a. Positive chronotropic effect

–**increases heart rate** by increasing the rate of phase 4 depolarization.

–More action potentials will occur because threshold potential is reached more quickly; thus, heart rate increases.

–The **mechanism is an increase in** I_f, the inward Na^+ current that is responsible for phase 4 depolarization in the SA node.

b. Positive dromotropic effect

–**increases conduction velocity through the AV node.**

–**decreases the PR interval.**

–Action potentials are conducted more quickly from atria to ventricles, and ventricular filling may be compromised.

III. Cardiac Muscle and Cardiac Output

A. Myocardial cell structure

1. Sarcomere

–is the contractile unit of the myocardial cell.

–is similar to that in skeletal muscle.

–runs from Z line to Z line.

–contains thick filaments (myosin) and thin filaments (actin, troponin, tropomyosin).

–As in skeletal muscle, shortening occurs according to a sliding filament model, which states that actin filaments slide along adjacent myosin filaments by breaking and re-forming cross-bridges between actin and myosin.

2. Intercalated discs

–occur at Z lines.

–maintain cell-to-cell cohesion.

3. Gap junctions

–are present at the intercalated discs.

–are **low-resistance paths** between cells, allowing for rapid electrical spread of action potentials.

–account for the observation that the heart behaves as an **electrical syncytium**.

4. Mitochondria

–are more numerous in cardiac muscle than in skeletal muscle.

5. T tubules

–are continuous with the cell membrane.

–invaginate the cells at the Z lines and **carry the action potential into the cell interior.**

–are well developed in the ventricles but poorly developed in atria.

6. Sarcoplasmic reticulum (SR)

–are small-diameter tubules in close proximity to the contractile elements.

–are the site of **storage and release of Ca^{2+} for excitation–contraction coupling**.

B. Steps in excitation–contraction coupling

1. The action potential spreads from the cell membrane into the T tubules.

2. During the plateau of the action potential, conductance to Ca^{2+} is increased, and Ca^{2+} enters the cell from the extracellular fluid (**inward Ca^{2+} current**).

3. This Ca^{2+} entry acts as a trigger for release of even more Ca^{2+} from the SR (**Ca^{2+}–triggered Ca^{2+} release**). The amount of Ca^{2+} released from the SR depends on the amount of Ca^{2+} previously stored and on the size of the inward current.

4. As a result of this Ca^{2+} release, **intracellular [Ca^{2+}] increases**.

5. Ca^{2+} binds to troponin C, removing inhibition of the actin–myosin interaction by troponin–tropomyosin.

6. Now, movement of the thick and thin filaments can occur and the myocardial cell contracts. **The magnitude of the tension developed is proportional to the rise in intracellular [Ca^{2+}].**

7. Relaxation occurs when Ca^{2+} is reaccumulated by the SR by an active Ca^{2+}–ATPase pump.

C. Contractility

–is the **intrinsic ability of the cardiac muscle to develop force at a given muscle length**.

–is also called **inotropism**.

–can be estimated by the **ejection fraction** (stroke volume/end-diastolic volume), which is **normally 0.55 (55%)**.

–Agents that produce an increase in contractility have a **positive inotropic** effect.

–Agents that produce a decrease in contractility have a **negative inotropic** effect.

1. Factors that increase contractility (positive inotropic effects)

a. Increased heart rate

–increases contractility because more action potentials per unit time means more Ca^{2+} entry into the myocardial cell, more Ca^{2+} released from the SR, and greater tension produced during contraction.

–**Examples** of the **effect of heart rate** are:

(1) Positive staircase or Bowditch staircase. Increased heart rate increases the strength of contraction in a stepwise fashion as the intracellular [Ca^{2+}] increases over several beats.

(2) Post-extrasystolic potentiation. The beat **following** an extrasystolic beat has increased strength of contraction because of the increased intracellular [Ca^{2+}].

b. Sympathetic stimulation (catecholamines) via β receptors (see Table 3-1)

–increases the strength of contraction by **two mechanisms:**

(1) It increases the entry of Ca^{2+} into the cell during the plateau of each cardiac action potential.

(2) It increases the activity of the Ca^{2+} pump of the SR (**phospholambam**); therefore, more Ca^{2+} will be accumulated and thus will be available for release in subsequent beats.

c. Cardiac glycosides (digitalis)

  —increase the strength of contraction by inhibiting Na^+–K^+ ATPase in the cardiac muscle cell membrane (Figure 3-5).

  —As a result, the intracellular $[Na^+]$ rises, diminishing the Na^+ gradient across the cell membrane.

  —Ca^{2+}–Na^+ exchange (a mechanism that extrudes Ca^{2+} from the cell) depends on the size of this Na^+ gradient and is diminished, causing a rise in the intracellular $[Ca^{2+}]$.

2. Factors that decrease contractility (negative inotropic effects) [see Table 3-1]

  —**Parasympathetic stimulation (ACh) via muscarinic receptors** decreases the strength of contraction in atria by decreasing Ca^{2+} entry into the cell during the plateau of the cardiac action potential (inward Ca^{2+} current).

D. Length–tension relationship (Figure 3-6)

  —describes the effect of ventricular cell length on the strength of contraction.

  —is similar to that in skeletal muscle.

1. Preload for ventricular muscle

  —is equivalent to **end-diastolic volume or venous filling pressure**.

  —When ventricular filling is increased, the ventricular muscle fibers are stretched (see Frank-Starling relationship, III D 5).

2. Afterload

  —is equivalent to **aortic pressure**.

  —is increased by increasing aortic pressure.

3. Sarcomere length

  —determines the maximum number of cross-bridges that can form.

  —determines maximum tension or strength of contraction.

4. Velocity of contraction at fixed muscle length

  —is maximal when there is no afterload.

  —is decreased by increasing afterload.

5. Frank-Starling relationship

  —describes the increase in cardiac output (or stroke volume) that occurs in response to an increase in venous pressure or end-diastolic volume (see Figure 3-6).

  —is based on the length–tension relationship; **increases in end-diastolic volume cause an increase in fiber length, which causes an increase in developed tension**.

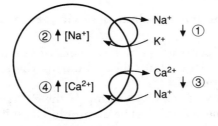

Myocardial cell

Figure 3-5. Stepwise explanation of how ouabain (digitalis) causes an increase in intracellular $[Ca^{2+}]$ and myocardial contractility.

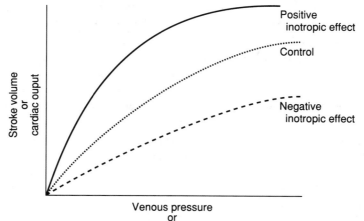

Figure 3-6. Frank-Starling relationship and the effect of positive and negative inotropic agents on it.

—is the mechanism that **matches cardiac output to venous return**.
—**Changes in contractility** shift the Frank-Starling curve upward (increased contractility) or downward (decreased contractility).

   a. Increases in contractility cause an increase in cardiac output for any level of venous pressure, right atrial pressure, or end-diastolic volume.

   b. Decreases in contractility cause a decrease in cardiac output for any level of venous pressure, right atrial pressure, or end-diastolic volume.

E. Ventricular pressure–volume loops (Figure 3-7)
   —are constructed by combining systolic and diastolic pressure curves.
   —The diastolic pressure curve is the relationship between diastolic pressure and diastolic volume in the ventricle.
   —The systolic pressure curve is the corresponding relationship between systolic pressure and systolic volume in the ventricle.
   —**A cycle of ventricular contraction, ejection, relaxation, and refilling can be visualized by combining the two curves into a pressure–volume loop.**

   1. 1 → 2 (isovolumetric contraction). Begin during diastole at point 1. The ventricle has been filled and its volume is about 140 ml (end-diastolic volume). The pressure is low because the ventricular muscle is relaxed.

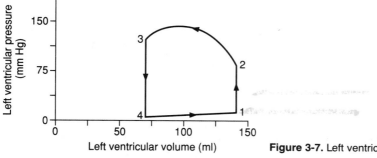

Figure 3-7. Left ventricular pressure–volume loop.

Upon excitation, the ventricle contracts and ventricular pressure increases. Because all valves are closed, no blood can be ejected from the ventricle (isovolumetric).

2. **2 → 3 (ventricular ejection).** The aortic valve opens at point 2 when pressure in the ventricle exceeds pressure in the aorta. Blood is ejected into the aorta, and the ventricular volume falls. The volume that is ejected is the **stroke volume**. Thus, stroke volume can be measured graphically by the width of the pressure–volume loop.

3. **3 → 4 (isovolumetric relaxation).** At point 3, the ventricle relaxes. When ventricular pressure falls to a value less than aortic pressure, the aortic valve closes. Because all valves are closed again, ventricular volume is constant (isovolumetric).

4. **4 → 1 (ventricular filling).** Once ventricular pressure falls below atrial pressure, the mitral valve opens and filling of the ventricle begins. During this phase, ventricular volume is increased back to about 140 ml (the end-diastolic volume).

F. Cardiac and vascular function curves (Figure 3-8)

–are simultaneous plots of cardiac output and venous return as a function of right atrial pressure or end-diastolic volume.

1. The cardiac output or cardiac function curve

–depicts the Frank-Starling relationship for the ventricle.
–shows that cardiac output increases as a function of end-diastolic volume—a consequence of the length–tension relationship in cardiac muscle fibers. **Remember that changes in end-diastolic volume are a major mechanism for altering cardiac output.**

2. The venous return or vascular function curve

–depicts the relationship between flow through the vascular system (or venous return) and the right atrial pressure.

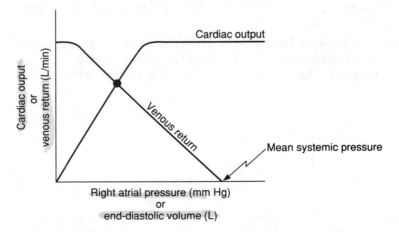

Figure 3-8. Simultaneous plots of the cardiac and vascular function curves. The curves cross at the equilibrium point for the cardiovascular system.

a. Mean systemic pressure

–equals right atrial pressure when there is "no flow" in the cardiovascular system.

–is measured when the heart is stopped experimentally. Under these conditions, cardiac output and venous return are zero, and pressure is equal throughout the entire cardiovascular system.

–is **increased by an increase in blood volume** and is **decreased by a decrease in blood volume.**

b. Slope of the venous return curve

–is **determined by the resistance of the vasculature**.

(1) A **clockwise rotation of the venous return curve** indicates a **decrease in TPR;** thus, when TPR is decreased for a constant right atrial pressure, there will be an increase in venous return (i.e., vasodilation of the arterioles "allows" blood to flow from the arteries to the veins and back to the heart).

(2) A **counterclockwise rotation of the venous return curve** indicates an **increase in TPR** (see Figure 3-11). Thus, when TPR is increased for a constant right atrial pressure, there will be a decrease in venous return to the heart.

3. Combining cardiac output and venous return curves

–The point where the two curves intersect is the **equilibrium or steady-state point**. Equilibrium occurs when cardiac output equals venous return.

–**Cardiac output can be changed** by altering the cardiac output curve, the venous return curve, or both. The superimposed curves can be used to predict the direction and magnitude of changes in cardiac output. **Examples** of such changes are:

a. Inotropic effects change the cardiac output curve.

(1) **Positive inotropic agents** (e.g., **digitalis**) produce increased contractility and increased cardiac output (Figure 3-9). The equilibrium point shifts to a higher cardiac output and a correspondingly lower right atrial pressure. Right atrial pressure is decreased because more blood is ejected from the heart on each beat (increased stroke volume).

(2) **Negative inotropic agents** produce decreased contractility and decreased cardiac output (not illustrated).

b. Changes in blood volume change the venous return curve.

(1) **Increases in blood volume** increase mean systemic pressure, shifting the venous return curve to the right in a parallel fashion (Figure 3-10). A new equilibrium point is established where **both cardiac output and right atrial pressure are increased**.

(2) **Decreases in blood volume** (e.g., hemorrhage; not illustrated) would have the opposite effect—decreased mean systemic pressure and a shift of the venous return curve to the left in a parallel fashion. A new equilibrium point is established where **both cardiac output and right atrial pressure are decreased.**

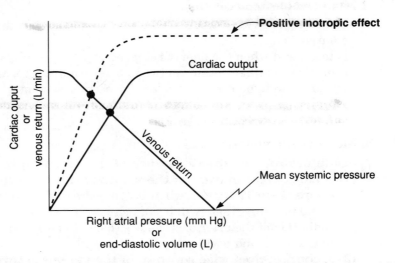

Figure 3-9. Effect of a positive inotropic agent on the cardiac function curve, cardiac output, and right atrial pressure.

c. Changes in TPR change both the cardiac output and venous return curves.

–Because changes in TPR alter both curves simultaneously, the responses are more complicated than those noted in the previous examples.

(1) Increasing TPR causes a decrease in both cardiac output and venous return (Figure 3-11).

(a) There is **counterclockwise rotation of the venous return curve.** Increased TPR results in decreased venous return as blood is retained on the arterial side.

(b) Downward shift of the cardiac output curve is caused by the increased aortic pressure (increased afterload) as the heart pumps against a higher pressure.

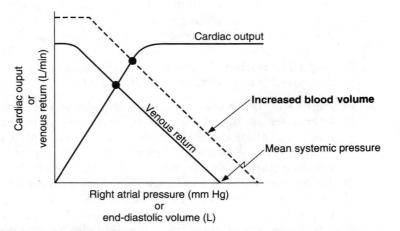

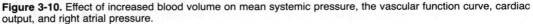

Figure 3-10. Effect of increased blood volume on mean systemic pressure, the vascular function curve, cardiac output, and right atrial pressure.

—As a result of these simultaneous changes, a new equilibrium point is established where **both cardiac output and venous return are decreased**.

(2) Decreasing TPR causes an increase in both cardiac output and venous return (not illustrated).

(a) There is a **clockwise rotation of the venous return curve**. Decreased TPR results in increased venous return as blood is allowed to flow back to the heart from the arterial side.

(b) Upward shift of the cardiac output curve is caused by the decreased aortic pressure (decreased afterload) as the heart pumps against a lower pressure.

—As a result of these simultaneous changes, a new equilibrium point is established where both **cardiac output and venous return are increased**.

G. Cardiac oxygen consumption

—is directly related to the amount of tension developed.

—is **increased by**:

1. Increased **afterload** (aortic pressure)

2. Increased **size of the heart** (law of LaPlace: tension is proportional to radius)

3. Increased **contractility**

4. Increased **heart rate**

—can be expressed by the following equation:

$$\textbf{Cardiac output} = \textbf{Stroke volume} \times \textbf{heart rate}$$

—The work done by the heart is **stroke work** and can be expressed by the following equation:

$$\textbf{Stroke work} = \textbf{Stroke volume} \times \textbf{aortic pressure}$$

—Fatty acids are the primary source of energy for stroke work.

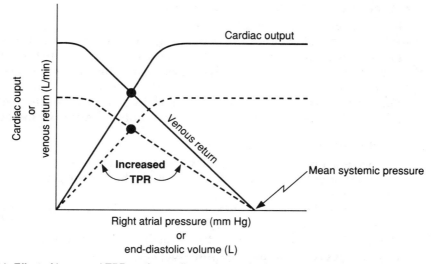

Figure 3-11. Effect of increased TPR on the cardiac and vascular function curves and on cardiac output.

H. Measurement of cardiac output by the Fick principle

−The Fick principle can be expressed by the following equation:

$$\text{Cardiac output} = \frac{O_2 \text{ consumption}}{[O_2]_{\text{pulmonary vein}} - [O_2]_{\text{pulmonary artery}}}$$

−The equation can be solved as follows:

1. **Oxygen consumption** for the whole body can be measured.
2. **Pulmonary vein** $[O_2]$ can be measured in a peripheral artery.
3. **Pulmonary artery** $[O_2]$ can be measured in mixed systemic venous blood.
   −**For example,** a man has a resting O_2 consumption of 250 ml/min, a peripheral arterial O_2 content of 0.20 ml O_2/ml of blood, and a mixed venous O_2 content of 0.15 ml O_2/ml of blood. What is his cardiac output?

$$\text{Cardiac output} = \frac{250 \text{ ml/min}}{(0.20 \text{ ml } O_2/\text{ml} - 0.15 \text{ ml } O_2/\text{ml})}$$

$$= 5000 \text{ ml/min or } 5.0 \text{ L/min}$$
(typical value for a 70-kg male)

IV. Cardiac Cycle (Figure 3-12)

−The figure shows the mechanical and electrical events of a single cardiac cycle. Seven phases are indicated by the vertical lines.

A. Atrial systole

−is preceded by the P wave.
−contributes to, but is not essential for, ventricular filling.
−The increase in atrial pressure (venous pressure) caused by atrial systole is the **a wave** in venous pulse.
−Filling of the ventricle by atrial systole causes the **fourth heart sound,** which is not audible in normal adults.

B. Isovolumetric ventricular contraction

−begins after the onset of the QRS wave.
−When ventricular pressure becomes greater than atrial pressure, the atrial–ventricular valves close; their closure corresponds to the **first heart sound**. Because the mitral valve closes before the tricuspid valve, the first heart sound may be split.
−Ventricular pressure rises isovolumetrically as a result of contraction, but no blood leaves the ventricle because the **aortic valve is closed.**

C. Rapid ventricular ejection

−Ventricular pressure reaches its maximum value.
−When ventricular pressure becomes greater than aortic pressure, the aortic valve opens.
−There is rapid ejection of blood into the aorta due to the pressure gradient between the ventricle and the aorta.
−Ventricular volume decreases dramatically since **most of the stroke volume is ejected during this phase**.
−Atrial filling begins.

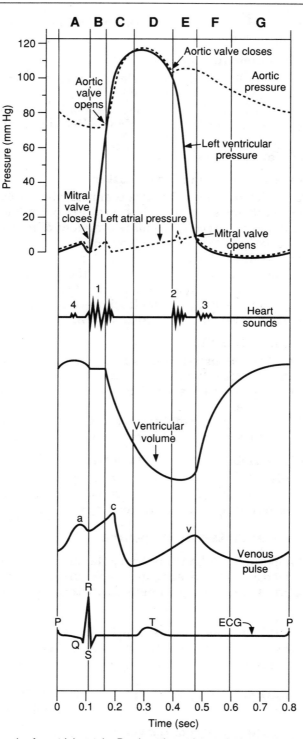

Figure 3-12. The cardiac cycle. *A* = atrial systole; *B* = isovolumetric ventricular contraction; *C* = rapid ventricular ejection; *D* = reduced ventricular ejection; *E* = isovolumetric ventricular relaxation; *F* = rapid ventricular filling; *G* = reduced ventricular filling.

—The onset of the T wave marks the end of the contraction and the end of rapid ventricular ejection.

D. Reduced ventricular ejection

—Ejection of blood from the ventricle continues but is slower.

—Ventricular pressure begins to fall.

—Aortic pressure also falls because run-off of blood from large arteries into smaller arteries is faster than the flow of blood from the ventricle into the aorta.

—Atrial filling continues.

E. Isovolumetric ventricular relaxation

—Repolarization of the ventricles is complete (the T wave).

—The aortic valve closes, followed by closure of the pulmonic valve; closure of the semilunar valves corresponds to the **second heart sound**. Inspiration causes splitting of the second heart sound.

—The atrial–ventricular valves remain closed.

—Ventricular pressure falls rapidly since the ventricle is now relaxed, and ventricular volume is constant because all valves are closed.

—The "blip" in the aortic pressure tracing occurs following closure of the aortic valve and is called the **dicrotic notch, or incisura**.

F. Rapid ventricular filling

—When ventricular pressure falls below atrial pressure, the mitral valve opens and the left ventricle begins to fill.

—Aortic pressure continues to fall because blood continues to run off into the smaller arteries.

—Rapid flow of blood from the atria into the ventricles causes the **third heart sound,** which is normal in children but in adults is associated with disease.

G. Reduced ventricular filling (diastasis)

—Ventricular filling continues but at a slower rate.

—The time for diastasis depends upon heart rate; increased heart rate decreases the time for ventricular refilling.

V. Regulation of Arterial Pressure

—The most important mechanisms for regulating arterial pressure are the **fast, neurally mediated baroreceptor mechanism** and the **slower, hormonally regulated renin–angiotensin–aldosterone mechanism**.

A. Baroreceptor reflex

—includes the **fast, neural mechanisms**.

—is responsible for the minute-to-minute regulation of arterial blood pressure.

—produces vasoconstrictor activity tonically, which accounts for **vasomotor tone**.

—**Baroreceptors** are stretch receptors located within the walls of the carotid sinus near the bifurcation of the common carotid arteries.

1. Steps in the baroreceptor reflex

a. An **increase in arterial pressure stretches the walls of the carotid sinus**.

—Because the baroreceptors are most sensitive to **changes in arterial pressure,** a rapidly increasing arterial pressure produces more response than a high but unchanging pressure.

—Additional baroreceptors in the **aortic arch** respond to increases, but not to decreases, in arterial pressure.

b. Stretch increases the firing rate of the carotid sinus nerve (Hering's nerve, cranial nerve IX), which carries information to the vasomotor center in the brainstem.

c. The set point for mean arterial blood pressure in the vasomotor center is about 100 mm Hg. Therefore, if mean arterial pressure is greater than 100 mm Hg, a series of autonomic responses are coordinated by the vasomotor center to reduce it.

d. The **responses of the vasomotor center to an increase in mean arterial blood pressure** are coordinated to decrease the arterial pressure back to 100 mm Hg. The responses are **increased parasympathetic (vagal) outflow to the heart and decreased sympathetic outflow to the heart and blood vessels**.

—The following four effects together bring the arterial pressure down toward normal:

(1) ↓ **heart rate,** resulting from increased parasympathetic tone and decreased sympathetic tone to the heart.

(2) ↓ **contractility,** resulting from decreased sympathetic tone to the heart. Together with the decrease in heart rate, cardiac output will be decreased and, as a result, arterial pressure will also be decreased.

(3) ↓ **vasoconstriction of arterioles,** resulting from the decreased sympathetic outflow. As a result, **TPR will decrease,** decreasing arterial pressure.

(4) ↓ **vasoconstriction of veins,** resulting from the decreased sympathetic outflow. As the veins relax, there will be an **increase in unstressed volume** and, as a result, mean systemic pressure will decrease.

e. This is a **negative feedback system**. As these mechanisms work together to decrease the mean arterial pressure back to normal, there will be decreased stretch on the carotid sinus baroreceptors and decreased signal to the vasomotor center.

2. Example of the baroreceptor reflex: response to acute blood loss (Figure 3-13)

B. Renin–angiotensin–aldosterone system

—is a slow, hormonal mechanism.

—is used in long-term blood pressure regulation by **adjustment of blood volume**.

—**Renin** is an enzyme that catalyzes the conversion of angiotensinogen to angiotensin I in the plasma.

—**Angiotensin I** is inactive.

—**Angiotensin II is physiologically active.**

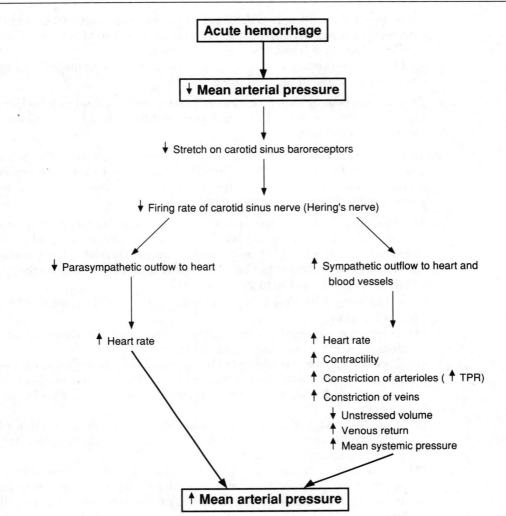

Figure 3-13. Role of the baroreceptor reflex in the cardiovascular response to hemorrhage.

—**Angiotensin II** is degraded by angiotensinases. One of the peptide fragments, **angiotensin III,** has some of the biologic activity of angiotensin II.

1. **Steps in the renin–angiotensin–aldosterone system**

   a. A **decrease in renal perfusion pressure** causes release of renin from the juxtaglomerular cells of the afferent arteriole.

   b. **Angiotensinogen is converted to angiotensin I** in plasma, catalyzed by renin.

   c. **Angiotensin I is converted to angiotensin II,** catalyzed by angiotensin-converting enzyme (ACE). The primary site of this reaction is the **lung.** Inhibitors of ACE can lower the blood pressure by blocking the production of angiotensin II.

d. Angiotensin II has two effects: **It stimulates release of aldosterone from the adrenal cortex,** and it causes **vasoconstriction of arterioles** (↑ TPR).

e. Aldosterone increases reabsorption of salt by the distal tubule of the kidney.

  —This action **is slow** because it requires the synthesis of new protein by the kidney.

  —Increased salt and water reabsorption **increases blood volume and mean arterial pressure**.

2. Example: response of the renin–angiotensin–aldosterone system to blood loss (Figure 3-14)

C. Other regulation of arterial blood pressure

1. Cerebral ischemia

a. When the brain is ischemic, the concentrations of CO_2 and H^+ in brain tissue increase.

b. Chemoreceptors in the vasomotor center respond by **increasing both sympathetic and parasympathetic outflow**.

  —Contractility and TPR are increased, but heart rate is slowed (due to the overriding parasympathetic influence); **mean arterial pressure can increase to life-threatening levels**.

  —Because of peripheral vasoconstriction, **flow to other organs (e.g., kidneys) is significantly reduced** in an attempt to preserve blood flow to the brain.

c. The **Cushing reaction** is an example. Increases in intracranial pressure cause compression of the cerebral blood vessels and cerebral ischemia. The response is as described above—elevation of mean arterial pressure (sympathetic increase in contractility and TPR), with simultaneous **reduction in heart rate** (parasympathetic).

2. Chemoreceptors in the carotid and aortic bodies

  —are located near the bifurcation of the common carotid arteries and along the aortic arch.

  —have very high rates of O_2 consumption and therefore are **very sensitive to hypoxia**. A decrease in mean arterial pressure causes a reduction in O_2 delivery to the chemoreceptors. In turn, information is sent to the vasomotor center to activate mechanisms to restore blood pressure.

3. Vasopressin (antidiuretic hormone; ADH)

  —is involved in blood pressure regulation in response to hemorrhage but not in minute-to-minute regulation of normal blood pressure.

  —Receptors in the atria respond to a decrease in volume (or pressure) and cause the release of vasopressin from the posterior pituitary.

  —Vasopressin has two effects that tend to increase blood pressure and return it to normal:

a. It is a **potent vasoconstrictor that increases TPR by activating V1 receptors on the arterioles.**

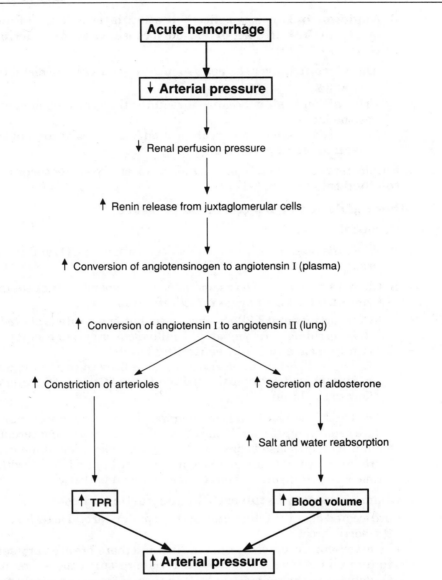

Figure 3-14. Role of the renin–angiotensin–aldosterone system in the cardiovascular response to hemorrhage.

    b. It increases reabsorption of water by the renal distal tubule and collecting ducts by activating V2 receptors.

  4. Atrial natriuretic peptide (ANP)
    –is **released from atria in response to an increase in atrial pressure.**
    –is a potent **inhibitor of vascular smooth muscle contraction,** causing dilation of arterioles and decreased TPR.
    –**causes excretion of increased amounts of salt and water** by the kidney, which reduces blood volume and attempts to bring arterial pressure down to normal.
    –**inhibits renin release.**

VI. Microcirculation and Lymph

A. Structure of capillary beds

- —Capillaries branch off from metarterioles; the junction is a band of smooth muscle called the **precapillary sphincter**.
- —True capillaries do not have smooth muscle; they consist of a single layer of **endothelial cells** surrounded by a basement membrane.
- —Clefts (pores) between the endothelial cells allow passage of water-soluble substances. The clefts represent a very small fraction of the surface area ($< 0.1\%$).
- —**Flow through capillaries is regulated by contraction and relaxation of the arterioles and the precapillary sphincters.**

B. Passage of substances across the capillary wall

1. Lipid-soluble substances

- —can cross the membranes of capillary endothelial cells by **simple diffusion**.
- —include O_2 and CO_2.

2. Small water-soluble substances

- —can cross via the water-filled clefts between the endothelial cells.
- —include **water, glucose, and amino acids**.
- —Generally, protein molecules are too large to pass freely through the clefts.
- —In the brain, the clefts between endothelial cells are exceptionally tight (**blood–brain barrier**).
- —In the liver and intestine, the clefts are exceptionally wide and allow passage of protein. These capillaries are called **sinusoids**.

3. Large water-soluble substances

- —can cross by **pinocytosis**.

C. Fluid exchange across capillaries

1. The Starling equation

$$J_v = K_f \left[(P_c - P_i) - (\pi_c - \pi_i) \right]$$

a. J_v is fluid flow.

- —If J_v is positive, then there is **net fluid movement out** of the capillary (**filtration**).
- —If J_v is negative, then there is **net fluid movement into** the capillary (**absorption**).

b. K_f is the filtration coefficient.

- —It is the hydraulic conductance of the capillary wall (permeability to water).

c. P_c is capillary hydrostatic pressure (mm Hg); increases in P_c favor filtration out of the capillary.

- —It is determined by the arterial and venous pressures and resistances.
- —An **increase in either arterial or venous pressure will increase P_c**; increases in venous pressure have the greater effect.
- —P_c is higher at the arteriolar end of the capillary than at the venous end (except in glomerular capillaries where it is nearly constant).

d. **P_i is interstitial fluid hydrostatic pressure (mm Hg); increases in P_i oppose filtration out of the capillary.**
   −It is normally close to 0 mm Hg (or negative).

e. **π_c is capillary oncotic or colloidosmotic pressure (mm Hg); increases in π_c oppose filtration out of the capillary.**
   −It is increased by increases in protein concentration in blood.
   −Small solutes do not contribute to π_c.

f. **π_i is interstitial fluid oncotic pressure (mm Hg); increases in π_i favor filtration out of the capillary.**
   −It is dependent upon the protein concentration of interstitial fluid.
   −Its value is normally quite low because little protein is filtered out of capillaries into interstitial fluid.

2. **Factors that increase filtration**

   a. ↑ P_c—caused by increased arterial pressure and by increased venous pressure

   b. ↓ P_i

   c. ↓ π_c—caused by decreased protein concentration in blood

   d. ↑ π_i—caused by inadequate lymphatic function

3. **Sample calculations using the Starling equation**

   a. **Example 1:** At the arteriolar end of a capillary, P_c is 30 mm Hg, π_c is 28 mm Hg, P_i is 0 mm Hg, and π_i is 4 mm Hg. Will filtration or absorption occur?

$$\text{Net pressure} = (30 - 0) - (28 - 4) \text{ mm Hg}$$
$$= +6 \text{ mm Hg}$$

   Thus, **filtration will occur**.

   b. **Example 2:** At the venous end of the same capillary, P_c has decreased to 16 mm Hg, π_c remains at 28 mm Hg, P_i is 0 mm Hg, and π_i is 4 mm Hg. Will filtration or absorption occur?

$$\text{Net pressure} = (16 - 0) - (28 - 4) \text{ mm Hg}$$
$$= -8 \text{ mm Hg}$$

   Thus, **absorption will occur**.

4. **Lymph**

   a. **Function of lymph**
      −Normally, filtration of fluid out of capillaries is slightly greater than absorption of fluid into the capillaries. The **excess filtered fluid is returned to the circulation via the lymph**.
      −Lymph also returns any filtered protein to the circulation.

   b. **Unidirectional flow of lymph**
      −One-way **flap valves** permit interstitial fluid to enter, but not leave, the lymph vessels.
      −Flow through larger lymphatic vessels is also **unidirectional, and is aided by one-way valves and skeletal muscle contraction**.

    c. Edema (Table 3-2)

        −forms when the volume of interstitial fluid exceeds the ability of the lymphatics to return it to the circulation.

        −can be caused by **excess filtration** or **blocked lymphatics**.

D. Endothelial-derived relaxing factor (EDRF)

    −is produced in the endothelial cells.

    −causes local **relaxation of vascular smooth muscle**.

    −Mechanism of action involves activation of guanylate cyclase and production of **cyclic guanosine monophosphate (GMP)**.

    −One form of EDRF may be **nitric oxide**.

    −Circulating ACh causes vasodilation by stimulation of EDRF production.

VII. Special Circulations (Coronary, Cerebral, Muscle, Skin)
[Table 3-3]

−Blood flow varies from one organ to the next.

−Blood flow to an organ is regulated by altering arteriolar resistance, and can be varied, depending on metabolic demands.

−Circulation to the lungs and kidneys is discussed in Chapters 4 and 5, respectively.

A. Local (intrinsic) control of blood flow

1. Examples of local control

a. Autoregulation

    −means that blood flow to an organ remains constant over a wide range of perfusion pressures.

    −Organs exhibiting autoregulation are the heart, brain, kidney, and skeletal muscle.

    −**For example,** if perfusion pressure to the heart is suddenly decreased, there will be a compensatory vasodilation in order to maintain a constant flow.

Table 3-2. Causes and Examples of Edema Formation

Cause	Examples
↑ P_c	Arteriolar dilation
	Venous constriction
	Increased venous pressure
	Heart failure
	Extracellular volume expansion
	Standing (edema in the dependent limbs)
↓ π_c	Decreased plasma protein concentration
	Severe liver disease (failure to synthesize proteins)
	Protein malnutrition
	Nephrotic syndrome (loss of protein in urine)
↑ K_f	Burn
	Inflammation (release of histamine; cytokines)

Table 3-3. Summary of Control of Special Circulations

Circulation* (% of resting cardiac output)	Local Metabolic Control	Vasoactive Metabolites	Sympathetic Control	Mechanical Effects
Coronary (5%)	Most important mechanism	Hypoxia Adenosine	Least important mechanism	Mechanical compression during systole
Cerebral (15%)	Most important mechanism	CO_2 H^+	Least important mechanism	Increases in intracranial pressure decrease cerebral blood flow
Muscle (20%)	Most important mechanism during exercise	Lactate K^+ Adenosine	Most important mechanism at rest (α receptor causes vasoconstriction; β receptor causes vasodilation)	Muscular activity causes temporary decrease in blood flow
Skin (5%)	Least important mechanism		Most important mechanism (temperature regulation)	
Pulmonary** (100%)	Most important mechanism	Hypoxia vasoconstricts	Least important mechanism	Lung inflation

*Renal blood flow (25% of resting cardiac output) is discussed in Chapter 5.
**Pulmonary blood flow is discussed in Chapter 4.

b. Active hyperemia

  - −Blood flow to an organ is proportional to its metabolic activity.
  - −**For example,** if metabolic activity in skeletal muscle increases as a result of strenuous exercise, blood flow to the muscle will increase proportionately to meet metabolic demands.

c. Reactive hyperemia

  - −is an increase in blood flow to an organ, and it occurs after a period of occlusion of flow.
  - −The longer the period of occlusion, the greater the increase in blood flow above preocclusion levels.

2. Mechanisms that explain local control of blood flow

a. Myogenic hypothesis

  - −**explains autoregulation** but not active hyperemia.
  - −**Vascular smooth muscle contracts when it is stretched.**
  - −**For example,** if perfusion pressure to an organ suddenly increases, the arteriolar smooth muscle will be stretched and will contract; the resulting vasoconstriction will maintain a constant flow.

b. Metabolic hypothesis

  - −Tissue supply of O_2 is matched to tissue demand for O_2.
  - −**Vasodilator metabolites** are produced as a result of tissue activity. These vasodilators are CO_2, H^+, K^+, **lactate, and adenosine.**

−**Examples of active hyperemia** include the following:

(1) If metabolic activity of a tissue increases, there will be an increased demand for O_2 and increased production of vasodilator metabolites. These will cause arteriolar vasodilation, increased blood flow, and increased O_2 delivery to the tissue to meet demand.

(2) If blood flow to an organ is increased as a result of a spontaneous increase in arterial pressure, then more O_2 will be provided for metabolic activity. At the same time, the increased flow will "wash out" vasodilator metabolites. As a result of this "wash-out," there will be arteriolar vasoconstriction and increased resistance, and blood flow will be decreased back to normal.

B. Hormonal (extrinsic) control of blood flow

1. Sympathetic innervation of vascular smooth muscle

−**Increases in sympathetic tone** cause vasoconstriction; **decreases in sympathetic tone** cause vasodilation.

−The density of innervation varies widely among tissues, with skin having the greatest innervation and coronary, pulmonary, and cerebral vessels having little innervation.

2. Other vasoactive hormones

a. Histamine

−causes **arteriolar vasodilation and venous vasoconstriction**.

−has an overall effect of **increasing P_c, increasing filtration out of the capillaries, and causing local edema**.

−is released in response to tissue trauma.

b. Bradykinin

−causes **arteriolar vasodilation and venous vasoconstriction**.

−produces, like histamine, increased filtration out of capillaries, and causes local edema.

c. Serotonin (5-hydroxytryptamine)

−causes arteriolar vasoconstriction and is released in response to blood vessel damage to help prevent blood loss.

−has been implicated in the vascular spasms of **migraine headaches**.

d. Prostaglandins

−**Prostacyclin** is a vasodilator in several vascular beds.

−**E-series prostaglandins** are vasodilators.

−**F-series prostaglandins** are vasoconstrictors.

−**Thromboxane A_2** is a vasoconstrictor.

C. Coronary circulation

−**is controlled almost entirely by local metabolic factors.**

−exhibits autoregulation and active and reactive hyperemia.

−The most important local metabolic factors are **hypoxia** and **adenosine**.

−**For example,** increases in contractility are accompanied by an increased demand for O_2. To meet this demand, there is vasodilation of coronary vessels and, accordingly, increased blood flow and O_2 delivery to the contracting heart muscle.

—**During systole,** mechanical compression of the coronary vessels reduces blood flow. Reactive hyperemia occurs after the period of occlusion to repay the O_2 debt.

—Sympathetic nerves play a minor role.

D. Cerebral circulation

—**is controlled almost entirely by local metabolic factors.**

—exhibits autoregulation and active and reactive hyperemia.

—The **most important local vasodilator substance in the cerebral circulation is CO_2** (or pH). If P_{CO_2} is increased (pH is decreased), then there is vasodilation of the cerebral arterioles to increase blood flow to the brain.

—Sympathetic nerves and other hormones play a minor role.

—Vasoactive substances in the general circulation have little or no effect on the cerebral circulation, because such substances are excluded by the blood–brain barrier.

E. Skeletal muscle

—is controlled by local metabolic factors and by the extrinsic sympathetic innervation of blood vessels in muscle.

—**At rest, sympathetic control of blood flow predominates.**

—During **exercise, local metabolic control overrides sympathetic control**.

1. Sympathetic innervation

—The arterioles of skeletal muscle are densely innervated by sympathetic fibers. The veins are also innervated but less densely.

—There are both α and β receptors on the blood vessels.

—**Stimulation of α receptors causes vasoconstriction.**

—**Stimulation of β receptors causes vasodilation.**

—Vasoconstriction of skeletal muscle arterioles is a major contributor to the TPR (because of the large mass of muscle).

2. Local metabolic control

—Blood flow in skeletal muscle exhibits autoregulation and active and reactive hyperemia.

—Demand for O_2 in skeletal muscle varies with activity level, and blood flow is regulated to meet demand.

—These local metabolic mechanisms dominate when demand is high, such as **during exercise**.

—The local vasodilator substances **lactate, adenosine, and potassium,** are implicated in control of muscle blood flow.

—Mechanical effects during exercise temporarily compress the arteries and decrease flow; reactive hyperemia increases flow during the postocclusion period to repay the O_2 debt.

F. Skin

—has **extensive sympathetic innervation, and blood flow is under extrinsic control**.

—**Temperature regulation** is the principal function of the sympathetic innervation (see Chapter 2). Increased ambient temperature leads to vasodilation, allowing dissipation of excess body heat.

–**Trauma** produces the "triple response" in skin—a red line, a red flare, and a wheal. The **wheal is edema that results from the local release of histamine, which increases capillary filtration**.

VIII. Integrative Functions of the Cardiovascular System: Gravity, Exercise, and Hemorrhage

–The responses to changes in gravitational forces, exercise, and hemorrhage demonstrate the integrative functions of the cardiovascular system.

A. Changes in gravitational forces (Table 3-4)

–The following **changes** will occur **when an individual moves from a supine position to a standing position:**

1. Upon standing, a significant volume of blood will pool in the extremities because of the high compliance of the veins. (Muscular activity prevents venous pooling of blood.)

2. As a result of **venous pooling** and increased venous pressure, P_c in the legs increases and fluid is filtered into the interstitium. If net filtration of fluid exceeds the ability of the lymphatics to return it to the circulation, **edema** will form.

3. As volume is lost from the vascular compartment, blood volume and venous return will decrease. As a result, both **stroke volume and cardiac output will also decrease** (Frank-Starling relationship).

4. Initially, arterial pressure will decrease because of the reduction in cardiac output. If cerebral blood pressure becomes low enough, fainting may occur.

5. **Compensatory mechanisms** will attempt to restore the blood pressure. The **carotid sinus baroreceptors** will respond to the decrease in arterial pressure by decreasing the firing rate of the carotid sinus nerve. A coordinated response from the vasomotor center will increase sympathetic outflow and decrease parasympathetic outflow. As a result, heart rate and TPR will increase, and blood pressure will return toward normal.

6. **Orthostatic hypotension** (fainting or lightheadedness upon standing) can occur in individuals whose baroreceptor reflex mechanism is impaired (e.g., individuals treated with sympatholytic agents).

B. Exercise (Table 3-5)

1. The central command (anticipation of exercise)

–originates in the motor cortex or from reflexes initiated in muscle proprioceptors when exercise is anticipated.

Table 3-4. Summary of Responses to Standing

Arterial blood pressure	↓ initially, then corrects
Heart rate	↑
Cardiac output	↓
Stroke volume	↓
TPR	↑
Central venous pressure	↓

Table 3-5. Summary of Effects of Exercise

Heart rate	↑ ↑
Stroke volume	↑
Cardiac output	↑ ↑
Arterial pressure	↑ (slight)
Pulse pressure	↑ (due to increased stroke volume)
TPR	↓ ↓ (due to vasodilation of skeletal muscle beds)
AV O_2 difference	↑ ↑ (due to increased O_2 consumption)

   a. Sympathetic outflow to the heart and blood vessels is increased. As a result, heart rate and contractility (stroke volume) are increased, and unstressed volume is decreased.

   b. Cardiac output is increased, primarily as a result of the increased heart rate and, to a lesser extent, the increased stroke volume.

   c. Venous return is increased as result of muscular activity. Increased venous return provides more blood for each stroke volume (Frank-Starling relationship).

   d. Resistance to the skin, splanchnic regions, kidneys, and inactive muscles is increased and flow to these organs is decreased.

2. Increased metabolic activity of skeletal muscle
   −**Vasodilator metabolites (lactate, K^+, and adenosine) accumulate** because of increased metabolism of the exercising muscle.
   −These metabolites cause dilation of arterioles and precapillary sphincters, thus increasing blood flow and the number of perfused capillaries (active hyperemia).
   −O_2 delivery to the muscle is increased, and diffusion distance for O_2 is decreased.
   −**This vasodilation accounts for the overall decrease in TPR** that occurs with exercise. Note that activation of the sympathetic nervous system alone (central command) would cause an increase in TPR.

C. Hemorrhage (Table 3-6)
   −**Compensatory responses to acute blood loss** are as follows:

1. A decrease in blood volume decreases mean systemic pressure. As a result, there is a **decrease in cardiac output and arterial pressure**.

Table 3-6. Summary of Responses to Hemorrhage

Heart rate	↑
TPR	↑
Contractility	↑
Unstressed volume	↓ (↑ stressed volume)
Renin	↑
Angiotensin II	↑
Aldosterone	↑
Circulating epinephrine and norepinephrine	↑
ADH	↑

2. The carotid sinus baroreceptors detect the decrease in arterial pressure. As a result, **sympathetic outflow to the heart and blood vessels is increased,** and **parasympathetic outflow to the heart is decreased,** producing:

    a. ↑ heart rate

    b. ↑ contractility

    c. ↑ TPR

    d. ↓ unstressed volume and ↑ stressed volume

    —Vasoconstriction occurs in skeletal, splanchnic, and cutaneous vascular beds. However, it does not occur in the coronary or cerebral vascular beds, ensuring that flow will be maintained to the heart and brain.

    —These responses attempt to restore the arterial blood pressure back to normal.

3. Chemoreceptors in the carotid and aortic bodies are very sensitive to hypoxia. They supplement the baroreceptor mechanism by increasing sympathetic outflow.

4. Cerebral ischemia (if any) causes an increase in PCO_2, which activates chemoreceptors in the vasomotor center to increase sympathetic outflow.

5. Arteriolar vasoconstriction causes a decrease in P$_c$. As a result, **capillary absorption is favored,** which increases circulating blood volume.

6. The adrenal medulla releases epinephrine and norepinephrine, which supplement the actions of the sympathetic nervous system.

7. The renin–angiotensin–aldosterone system is activated by the decrease in renal perfusion pressure. Because angiotensin II is a potent vasoconstrictor, it reinforces the stimulatory effect of the sympathetic nervous system on TPR. Aldosterone increases salt and water reabsorption in the kidney, increasing the circulating blood volume.

8. ADH is released when atrial receptors detect the decrease in blood volume. ADH causes both vasoconstriction and increased water reabsorption in the kidney, both of which will tend to raise blood pressure.

Review Test

Directions: Each of the numbered items or incomplete statements in this section is followed by answers or by completions of the statement. Select the **one** lettered answer or completion that is **best** in each case.

1. Arteriography of a patient's left renal artery shows narrowing of the artery's radius by 50%. What is the expected change in blood flow through the stenotic artery?

(A) Decrease to ½
(B) Decrease to ¼
(C) Decrease to ⅛
(D) Decrease to 1/16
(E) Unchanged

2. When a person moves from a supine position to a standing position, which of the following compensatory changes occurs?

(A) Decreased heart rate
(B) Increased contractility
(C) Decreased TPR
(D) Decreased cardiac output
(E) Increased PR intervals

3. A person's ECG has no P wave, but a normal QRS and T wave. His pacemaker is located in the

(B) SA node
(B) AV node
(C) bundle of His
(D) Purkinje system
(E) ventricular muscle

Questions 4 and 5

An ECG on a person shows ventricular extra-systoles.

4. The extrasystolic beat would produce

(A) an increased pulse pressure because contractility is increased
(B) an increased pulse pressure because heart rate is increased
(C) a decreased pulse pressure because ventricular filling time is increased
(D) a decreased pulse pressure because stroke volume is decreased
(E) a decreased pulse pressure because PR interval is increased

5. The next "normal" ventricular contraction following the extrasystole would produce

(A) an increased pulse pressure because contractility of the ventricle is increased
(B) an increased pulse pressure because TPR is decreased
(C) an increased pulse pressure because compliance of the veins is decreased
(D) a decreased pulse pressure because contractility of the ventricle is increased
(E) a decreased pulse pressure because TPR is decreased

6. An increase in contractility is demonstrated on a Frank-Starling diagram by

(A) increased cardiac output for a given end-diastolic volume
(B) increased cardiac output for a given end-systolic volume
(C) decreased cardiac output for a given end-diastolic volume
(D) decreased cardiac output for a given end-systolic volume

Questions 7–10

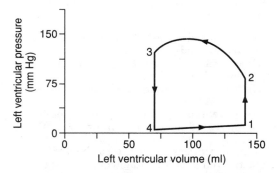

7. On the graph showing left ventricular volume and pressure, isovolumetric contraction occurs from point

(A) 4 → 1
(B) 1 → 2
(C) 2 → 3
(D) 3 → 4

8. The aortic valve closes at point

(A) 1
(B) 2
(C) 3
(D) 4

9. The first heart sound corresponds to point

(A) 1
(B) 2
(C) 3
(D) 4

10. If the heart rate is 70 beats/min, then cardiac output from this ventricle is closest to

(A) 3.45 L/min
(B) 4.55 L/min
(C) 5.25 L/min
(D) 8.0 L/min
(E) 9.85 L/min

Questions 11 and 12

In a capillary, P_c is 30 mm Hg, P_i is -2 mm Hg, π_c is 25 mm Hg, and π_i is 2 mm Hg.

11. What is the direction of fluid movement and the net driving force?

(A) Absorption; 6 mm Hg
(B) Absorption; 9 mm Hg
(C) Filtration; 6 mm Hg
(D) Filtration; 9 mm Hg
(E) There is no net fluid movement

12. If K_f is 0.5 ml/min/mm Hg, what is the rate of water flow across the capillary wall?

(A) 0.06 ml/min
(B) 0.45 ml/min
(C) 4.5 ml/min
(D) 9.0 ml/min
(E) 18.0 ml/min

13. Blood flow to which organ is NOT controlled primarily by metabolic factors (local metabolites)?

(A) Skin
(B) Lungs
(C) Heart
(D) Brain
(E) Skeletal muscle during exercise

14. All of the following parameters are increased during moderate exercise EXCEPT

(A) arteriovenous difference for O_2
(B) heart rate
(C) cardiac output
(D) pulse pressure
(E) TPR

15. The tendency for blood flow to be turbulent is increased by

(A) increased viscosity
(B) increased hematocrit
(C) partial occlusion of a blood vessel
(D) decreased velocity of blood flow

16. A patient experiences orthostatic hypotension after a sympathectomy. The explanation for this is

(A) an exaggerated response of the renin–angiotensin–aldosterone system
(B) a suppressed response of the renin–angiotensin–aldosterone system
(C) an exaggerated response of the baroreceptor mechanism
(D) a suppressed response of the baroreceptor mechanism

17. The ventricles are depolarizing or are depolarized during all of the following ECG waves, segments, or intervals EXCEPT

(A) QRS complex
(B) QT interval
(C) ST segment
(D) P wave

18. The ventricles are completely depolarized during which isoelectric portion of the ECG?

(A) PR interval
(B) QRS complex
(C) QT interval
(D) ST segment
(E) T wave

19. All of the following agents or changes have a positive inotropic effect on the heart EXCEPT

(A) increased heart rate
(B) sympathetic stimulation
(C) norepinephrine
(D) acetylcholine (ACh)
(E) cardiac glycosides

20. The change indicated by the dashed lines on the cardiac output/venous return curves shows

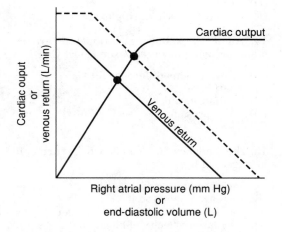

(A) decreased cardiac output in the "new" steady state
(B) decreased venous return in the "new" steady state
(C) increased mean systemic pressure
(D) decreased blood volume
(E) increased myocardial contractility

21. A person's ECG shows two P waves preceding each QRS complex. The interpretation of this pattern is

(A) decreased firing rate of the pacemaker in the SA node
(B) decreased firing rate of the pacemaker in the AV node
(C) increased firing rate of the pacemaker in the SA node
(D) decreased conduction through the AV node
(E) increased conduction through the AV node

22. An acute decrease in arterial blood pressure elicits which of the following compensatory changes?

(A) Decreased firing rate of the carotid sinus nerve
(B) Increased parasympathetic outflow to the heart
(C) Decreased heart rate
(D) Decreased contractility
(E) Decreased mean systemic pressure

23. The tendency to form edema will be increased by

(A) arteriolar constriction
(B) increased venous pressure
(C) increased plasma protein concentration
(D) dehydration

24. "Splitting" of the second heart sound occurs because

(A) the aortic valve closes before the pulmonic valve
(B) the pulmonic valve closes before the aortic valve
(C) the mitral valve closes before the tricuspid valve
(D) the tricuspid valve closes before the mitral valve
(E) filling of the ventricles has fast and slow components

Questions 25 and 26

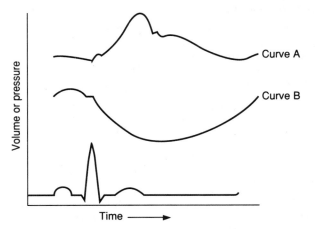

25. Curve A represents

(A) aortic pressure
(B) ventricular pressure
(C) atrial pressure
(D) ventricular volume

26. Curve B represents

(A) aortic pressure
(B) ventricular pressure
(C) atrial pressure
(D) ventricular volume

27. An increase in arteriolar resistance without a change in any other component of the cardiovascular system will produce

(A) a decrease in TPR
(B) an increase in capillary filtration
(C) an increase in arterial pressure
(D) a decrease in afterload

28. The following measurements were obtained in a male patient:

Heart rate $= 70$ beats/min
Pulmonary vein $[O_2] = 0.24$ ml O_2/ml
Pulmonary artery $[O_2] = 0.16$ ml O_2/ml
Whole body O_2 consumption $= 500$ ml/min

What is this patient's cardiac output?

(A) 1.65 L/min
(B) 4.55 L/min
(C) 5.0 L/min
(D) 6.25 L/min
(E) 8.0 L/min

Questions 29 and 30

29. The dashed line on the diagram illustrates the effect of

(A) increased TPR
(B) increased blood volume
(C) increased contractility
(D) negative inotropic agent
(E) increased mean systemic pressure

30. The X-axis on the diagram could have been labeled

(A) end-systolic volume
(B) end-diastolic volume
(C) pulse pressure
(D) mean systemic pressure
(E) heart rate

31. The greatest pressure drop in the circulation occurs across the arterioles because

(A) they have the greatest surface area
(B) they have the greatest cross-sectional area
(C) the velocity of blood flow through them is the highest
(D) the velocity of blood flow through them is the lowest
(E) they have the greatest resistance

32. Pulse pressure is

(A) the highest pressure measured in the arteries
(B) the lowest pressure measured in the arteries
(C) measured only during diastole
(D) determined by stroke volume
(E) decreased when capacitance of the arteries decreases

33. In the SA node, phase 4 depolarization (pacemaker potential) is attributable to

(A) an increase in K^+ conductance
(B) an increase in Na^+ conductance
(C) a decrease in Cl^- conductance
(D) a decrease in Ca^{2+} conductance
(E) simultaneous increases in K^+ and Cl^- conductances

34. A decrease in blood volume would be expected to cause an increase in all of the following EXCEPT

(A) renin secretion
(B) circulating levels of angiotensin II
(C) circulating levels of aldosterone
(D) renal perfusion pressure
(E) renal Na^+ reabsorption

35. All of following agents are released or secreted following a hemorrhage EXCEPT

(A) aldosterone
(B) angiotensin I
(C) angiotensin II
(D) ADH
(E) atrial natriuretic peptide

36. The low-resistance pathways between myocardial cells that allow for spread of action potentials are the

(A) gap junctions
(B) T tubules
(C) sarcoplasmic reticulum
(D) intercalated discs
(E) mitochondria

37. Myocardial contractility is best correlated with the intracellular concentration of

(A) Na^+
(B) K^+
(C) Ca^{2+}
(D) Cl^-
(E) Mg^{2+}

38. Increases in all of the following will cause an increase in myocardial O_2 consumption EXCEPT

(A) aortic pressure
(B) heart rate
(C) contractility
(D) size of the heart
(E) influx of Na^+ during the upstroke of the action potential

39. Each of the following substances crosses capillary walls primarily through H_2O-filled clefts between the endothelial cells EXCEPT

(A) O_2
(B) Na^+
(C) water
(D) glucose
(E) alanine

40. Which of the following is an effect of histamine?

(A) Decreased capillary filtration
(B) Vasodilation of arterioles
(C) Vasodilation of veins
(D) Decreased P_c
(E) Interaction with muscarinic receptors on the blood vessels

41. CO_2 regulates blood flow to which one of the following organs?

(A) Heart
(B) Skin
(C) Brain
(D) Skeletal muscle at rest
(E) Skeletal muscle during exercise

42. Cardiac output of the right heart is what percentage of that of the left heart?

(A) 25%
(B) 50%
(C) 75%
(D) 100%
(E) 125%

43. The physiologic function of the relatively slow conduction through the AV node is to allow sufficient time for

(A) run-off of blood from the aorta to the arteries
(B) venous return to the atria
(C) filling of the ventricles
(D) contraction of the ventricles
(E) repolarization of the ventricles

Directions: Each group of items in this section consists of lettered options followed by a set of numbered items. For each item, select the **one** lettered option that is most closely associated with it. Each lettered option may be selected once, more than once, or not at all.

Questions 44–46

Match each numbered phenomenon below with the appropriate phase of the ventricular action potential shown in the graph.

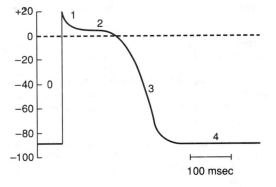

100 msec

(A) Phase 0
(B) Phase 1
(C) Phase 2
(D) Phase 3
(E) Phase 4

44. Phase of the ventricular action potential in which the membrane potential is closest to the K^+ equilibrium potential

45. Phase of the ventricular action potential that has the highest conductance to Ca^{2+}

46. Phase of the ventricular action potential that coincides with diastole

Questions 47 and 48

(A) α Receptors
(B) β_1 Receptors
(C) β_2 Receptors
(D) Muscarinic receptors

47. Mediate constriction of arteriolar smooth muscle

48. Mediate slowing of the heart

Answers and Explanations

1–D. If the radius of the artery decreased by 50% (½), then resistance would increase by 2^4 or 16 ($R = 8\eta l/\pi r^4$). Because blood flow is inversely proportional to resistance ($Q = \Delta P/R$), flow will decrease to $\frac{1}{16}$ the original value.

2–B. When a person moves to a standing position, blood pools in the leg veins, causing a decrease in venous return to the heart, decreased cardiac output, and decreased arterial pressure. The baroreceptors will detect the decrease in arterial pressure, and the vasomotor center will be activated to increase sympathetic outflow and decrease parasympathetic outflow. Heart rate increases (and, as a result, PR interval will decrease); contractility and TPR will increase. Because both heart rate and contractility are increased, cardiac output will increase toward normal.

3–B. The absent P wave indicates that the atrium is not depolarizing and so the pacemaker must not reside in the SA node. Since the QRS and T waves are normal, then depolarization and repolarization of the ventricle must be proceeding in the normal sequence. This can happen if the pacemaker is located in the AV node. If the pacemaker were located in the bundle of His or in the Purkinje system, the ventricle would activate in an abnormal sequence (depending on the location of the pacemaker) and the QRS would have an abnormal configuration. The ventricular muscle does not have pacemaker properties.

4–D. On the extrasystolic beat, pulse pressure will decrease because there is inadequate ventricular filling time—the ventricle beats "too soon." As a result, stroke volume will be decreased.

5–A. The post-extrasystolic contraction will produce increased pulse pressure because contractility is increased. Extra Ca^{2+} enters the cell during the extrasystolic beat. Contractility is directly related to the amount of intracellular Ca^{2+} available for binding to troponin C and for removing inhibition of the actin–myosin interaction.

6–A. An increase in contractility produces an increase in cardiac output for a given end-diastolic volume or pressure. The Frank-Starling relationship demonstrates the matching of cardiac output (what leaves the heart) to venous return (what returns to the heart). An increase in contractility (positive inotropic effect) will shift the curve upward.

7–B. Isovolumetric contraction occurs during ventricular systole, before the aortic valve opens. Ventricular pressure increases, but volume remains constant because blood cannot be ejected into the aorta against a closed valve.

8–C. Closure of the aortic valve occurs once ejection of blood from the ventricle has occurred and left ventricular pressure has fallen below aortic pressure.

9–A. The first heart sound corresponds to closure of the atrial–ventricular valves. Prior to their closure, the ventricle was filling (phase $4 \rightarrow 1$). After their closure, isovolumetric contraction begins and ventricular pressure rises (phase $1 \rightarrow 2$).

10–C. Stroke volume is the volume ejected from the ventricle and is represented on the pressure-volume loop as phase $2 \rightarrow 3$; end-diastolic volume is about 140 ml and end-systolic volume is about 65 ml; the difference, stroke volume, is 75 ml. Cardiac output is calculated as stroke volume × heart rate or 75 ml × 70/minute = 5250 ml/min or 5.25 L/min.

11–D. The net driving force can be calculated using the Starling equation.

$$\text{Net pressure} = (P_c - P_i) - (\pi_c - \pi_i)$$
$$= [(30 - (-2) - (25 - 2)] \text{ mm Hg}$$
$$= 32 \text{ mm Hg} - 23 \text{ mm Hg}$$
$$= + 9 \text{ mm Hg}$$

Since the net pressure is + , filtration out of the capillary will occur.

12–C. K_f is the filtration coefficient for the capillary and describes the intrinsic water permeability.

$$\text{Water flow} = K_f \times \text{net pressure}$$
$$= 0.5 \text{ ml/min/mm Hg} \times 9 \text{ mm Hg}$$
$$= 4.5 \text{ ml/min}$$

13–A. Circulation of the skin is controlled primarily by the sympathetic nervous innervation. Coronary, cerebral, and pulmonary circulations are primarily regulated by local metabolic factors. Skeletal muscle circulation is regulated by metabolic factors (local metabolites) during exercise, although at rest it is controlled by the sympathetic innervation.

14–E. In anticipation of exercise, the central command increases sympathetic outflow to the heart and blood vessels, causing an increase in heart rate and contractility. Venous return is increased by muscular activity and contributes to an increase in cardiac output by the Frank-Starling mechanism. Pulse pressure will be increased because stroke volume is increased. While increased sympathetic outflow to the blood vessels might be expected to increase TPR, it does not because there is an overriding vasodilation of skeletal muscle arterioles due to the buildup of vasodilator metabolites (lactate, K^+, adenosine). Because this vasodilation improves the delivery of O_2, more O_2 can be extracted and utilized by the contracting muscle.

15–C. Turbulent flow is predicted when the Reynold's number is increased. Factors that increase the Reynold's number and produce turbulent flow are decreased viscosity (hematocrit) and increased velocity. Partial occlusion of a blood vessel increases Reynold's number (and turbulence) because the decrease in cross-sectional area results in increased blood velocity (v = Q/A).

16–D. Orthostatic hypotension is a decrease in arterial pressure when going from a supine to a standing position. A person with a normal baroreceptor mechanism responds to a decrease in arterial pressure through the vasomotor center by increasing sympathetic outflow and decreasing parasympathetic outflow. The sympathetic component helps to restore blood pressure by increasing heart rate, contractility, TPR, and mean systemic pressure. In a patient who has undergone a sympathectomy, the sympathetic component of the baroreceptor mechanism is absent.

17–D. Depolarization of the ventricles is represented on the ECG by the QRS complex, which is included in the QT interval. The ST segment represents the period during which the entire ventricle is depolarized. The T wave represents ventricular repolarization. During the P wave (atrial depolarization), the ventricles have not yet been activated.

18–D. The PR interval and ST segments are the only portions of the ECG that are isoelectric. The PR interval is the period of AV node conduction and the ventricles are not yet depolarized. The ST segment is the isoelectric period when the entire ventricle is depolarized.

19–D. A positive inotropic effect is one that increases myocardial contractility. Contractility is the ability to develop tension at a fixed muscle length. Factors that increase contractility are those that increase the intracellular $[Ca^{2+}]$. Increasing heart rate increases intracellular $[Ca^{2+}]$ because more Ca^{2+} ions enter the cell during the plateau of each action potential. Sympathetic stimulation and norepinephrine increase intracellular $[Ca^{2+}]$ by increasing entry during the plateau and increasing the storage of Ca^{2+} by the SR (for later release). Cardiac glycosides increase intracellular $[Ca^{2+}]$ by inhibiting the Na^+–K^+ pump, thereby inhibiting Ca^{2+}–Na^+ exchange (a mechanism that pumps Ca^{2+} out of the cell). ACh has a negative inotropic effect on atria.

20–C. The shift in the venous return curve to the right is consistent with an increase in blood volume and, as a consequence, mean systemic pressure. Both cardiac output and venous return are increased in the new steady state (and are equal to each other). Contractility is unaffected.

21–D. A pattern of two P waves preceding each QRS complex indicates that only every other P wave is conducted through the AV node to the ventricle. Thus, conduction velocity through the AV node must be decreased.

22–A. A decrease in blood pressure causes decreased stretch of the carotid sinus baroreceptors and decreased firing of the carotid sinus nerve. In an attempt to restore the blood pressure, the parasympathetic outflow to the heart is decreased and sympathetic outflow is increased. As a result, heart rate and contractility will be increased. Mean systemic pressure will increase because of increased sympathetic tone of the veins (and a shift of blood to the arteries).

23–B. Edema forms when more fluid is filtered out of the capillaries than can be returned to the circulation by the lymphatics. Filtration is increased by changes that increase P_c or decrease π_c. Arteriolar constriction would decrease P_c and decrease filtration. Dehydration would increase plasma protein concentration (by hemoconcentration) and thereby increase π_c and decrease filtration. Increased venous pressure would increase P_c and increase filtration.

24–A. The second heart sound results from closure of the aortic and pulmonic valves. Because the aortic valve closes before the pulmonic valve, the sound can be split by inspiration.

25–A. The ECG tracing serves as a reference. The QRS complex marks ventricular depolarization, followed immediately by ventricular contraction. Aortic pressure increases steeply following QRS, as blood is ejected from the ventricles. After reaching peak pressure, aortic pressure falls as blood runs off into the arteries. The characteristic dicrotic notch (blip in the aortic pressure curve) appears when the aortic valve closes. Aortic pressure continues to fall as blood flows out of the aorta.

26–D. Ventricular volume increases slightly with atrial systole (P wave), is constant during isovolumetric contraction (QRS), and then falls dramatically following the QRS when blood is ejected from the ventricle.

27–C. An increase in arteriolar resistance will increase TPR. Arterial pressure = cardiac output × TPR, so arterial pressure will also increase. Capillary filtration would decrease if the arterioles were constricted, because P_c would be decreased. Afterload of the heart would be increased by an increase in TPR.

28–D. Cardiac output can be calculated by the Fick principle if whole body O_2 consumption and $[O_2]$ in the pulmonary artery and pulmonary vein are measured. Mixed venous blood could substitute for a pulmonary artery sample, and peripheral arterial blood could substitute for a pulmonary vein sample. Heart rate is not needed.

$$\text{Cardiac output} = \frac{500 \text{ ml/min}}{0.24 \text{ ml } O_2/\text{ml} - 0.16 \text{ ml } O_2/\text{ml}}$$
$$= 6250 \text{ ml/min or } 6.25 \text{ L/min}$$

29–C. An upward shift of the cardiac output curve is consistent with an increase in myocardial contractility; for any right atrial pressure (sarcomere length), the force of contraction is increased. Such a change causes an increase in stroke volume and cardiac output. Increased blood volume and increased mean systemic pressure are related and would cause a rightward shift in the venous return curve. A negative inotropic agent would cause a decrease in contractility and a downward shift of the cardiac output curve.

30–B. End-diastolic volume and right atrial pressure can be used interchangeably.

31–E. The fall in pressure at any level of the cardiovascular system is caused by resistance ($\Delta P = Q \times R$). The greater the resistance, the greater the fall in pressure. The arterioles are the site of highest resistance in the vasculature. The arterioles do not have the greatest surface area or cross-sectional area (the capillaries do). Velocity of blood flow is neither the highest nor the lowest in the cardiovascular system.

32–D. Pulse pressure is the difference between the highest (systolic) and lowest (diastolic) arterial pressures. It reflects the volume ejected by the left ventricle (stroke volume). Pulse pressure increases when the capacitance of the arteries decreases, such as with aging.

33–B. Phase 4 depolarization is responsible for the pacemaker property of SA nodal cells. It is caused by an increase in Na^+ conductance and an inward Na^+ current (I_f), which depolarizes the cell membrane.

34–D. A decrease in blood volume would decrease renal perfusion pressure and would cause increased secretion of renin, increased circulating levels of angiotensin II and aldosterone, and increased Na^+ reabsorption.

35–E. Angiotensin I, angiotensin II, and aldosterone are increased in response to a decrease in renal perfusion pressure. ADH is released when atrial receptors detect a decrease in blood volume. Atrial natriuretic peptide is released in response to an increase in atrial pressure and an increase in its secretion would not be anticipated following blood loss.

36–A. The gap junctions occur at the intercalated discs between cells and are low-resistance sites of current spread.

37–C. Contractility is affected by the intracellular $[Ca^{2+}]$, which is regulated by Ca^{2+} entry across the cell membrane during the plateau of the action potential and by uptake into and release from the SR. Ca^{2+} binds to troponin C and removes the inhibition of actin–myosin interaction, allowing contraction (shortening) to occur.

38–E. Myocardial O_2 consumption is determined by the amount of tension developed by the heart. It increases when there are increases in aortic pressure (increased afterload), heart rate or stroke volume (which increase cardiac output), or when the size of the heart is increased ($T = P \times r$). Influx of Na^+ ions during an action potential is a purely passive process, driven by the electrochemical driving forces on Na^+ ions. Of course, maintenance of the inwardly directed Na^+ gradient over the long term requires the Na^+–K^+ pump, which is energized by ATP.

39–A. Because O_2 is lipophilic, it crosses capillary walls primarily by diffusion through the endothelial cell membranes. The other substances are water-soluble; they cannot cross through the lipid component of the cell membrane and are restricted to the water-filled clefts or pores between the cells.

40–B. Histamine causes vasodilation of arterioles, which increases P_c and capillary filtration. It also causes constriction of veins, which contributes to the increase in P_c. ACh interacts with muscarinic receptors (although these are not present on vascular smooth muscle).

41–C. Blood flow to the brain is autoregulated by the P_{CO_2}. If metabolism increases (or pressure decreases), the P_{CO_2} will increase and cause vasodilation. Blood flow to the heart and skeletal muscle during exercise is also regulated metabolically, but adenosine and hypoxia are the most important vasodilators for the heart; adenosine, lactate, and K^+ are most important for exercising skeletal muscle. Blood flow to the skin is regulated by the sympathetic nervous system.

42–D. Cardiac output of the left and right sides of the heart is equal. Blood ejected from the left heart to the systemic circulation must be oxygenated by passage through the pulmonary circulation.

43–C. The AV delay (which corresponds to the PR interval) allows time for filling of the ventricles from the atria. If the ventricles contracted before they were filled, stroke volume would be decreased.

44–E. Phase 4 is the resting membrane potential. Because the conductance to K^+ is highest, the membrane potential approaches the equilibrium potential for K^+.

45–C. Phase 2 is the plateau of the ventricular action potential. During this period, the conductance to Ca^{2+} increases transiently. Ca^{2+} that enters the cell during the plateau is the trigger that releases more Ca^{2+} from the SR for the contraction.

46–E. Phase 4 is electrical diastole.

47–A. The α receptors for norepinephrine are excitatory on vascular smooth muscle and cause constriction. There are β_2 receptors on the arterioles of skeletal muscle, but they produce vasodilation.

48–D. ACh causes slowing of the heart via muscarinic receptors in the SA node.

4
Respiratory Physiology

I. Lung Volumes and Capacities

A. Lung volumes (Figure 4-1)

1. Tidal volume

−is the volume inspired or expired with each normal breath.

2. Inspiratory reserve volume

−is the volume that can be inspired over and above the tidal volume.
−is used during exercise.

3. Expiratory reserve volume

−is the volume that can be expired after the expiration of a tidal volume.

4. Residual volume

−is the volume that remains in the lungs after a maximal expiration.
−**cannot be measured by spirometry.**

5. Dead space

   a. Anatomic dead space (Fowler method)

     −is the volume of the conducting airways.
     −is normally about 150 ml.

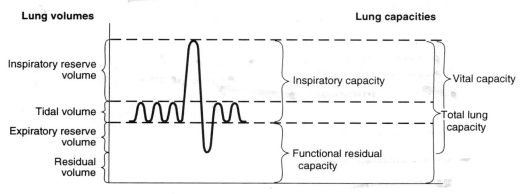

Figure 4-1. Lung volumes and capacities.

b. Physiologic dead space (Bohr method)

  –is a functional measurement.
  –is defined as the volume of the lung that does not eliminate CO_2.
  –may be greater than the anatomic dead space in lung disease where there are ventilation/perfusion (V/Q) inequalities.

6. Ventilation rate

a. Minute ventilation is expressed as follows:

  Minute ventilation = tidal volume × breaths/minute

b. Alveolar ventilation is expressed as follows:

  Alveolar ventilation = (tidal volume – dead space) × breaths/minute

  ↳ subtract out portion not participating in gas exchange

B. Lung capacities (see Figure 4-1)

1. Inspiratory capacity

  –is the sum of tidal volume and inspiratory reserve volume.

2. Functional residual capacity (FRC)

  –is the sum of expiratory reserve volume and residual volume.
  –is the volume remaining in the lungs after a tidal volume is expired.
  –includes the residual volume, so it **cannot be measured by spirometry**.

3. Vital capacity

  –is the sum of tidal volume, inspiratory reserve volume, and expiratory reserve volume.
  –is the volume that is expired after a maximal inspiration.

4. Total lung capacity

  –is the sum of all four lung volumes.
  –is the volume in the lungs after a maximal inspiration.
  –includes residual volume, so it **cannot be measured by spirometry**.

C. Forced expiratory volume (FEV$_1$)

  –is the volume of air that can be expired in one second following a maximal inspiration.
  –is **normally 80% of the forced vital capacity (FVC)** and is expressed as:

  $$FEV_1/FVC = 0.8$$

  –In restrictive lung disease such as **fibrosis,** both FEV_1 and FVC are reduced.
  –In obstructive lung disease such as **asthma,** FEV_1 is reduced more than FVC so that FEV_1/FVC is decreased.

II. Mechanics of Breathing

A. Muscles of inspiration

1. Diaphragm

  –is the **most important** muscle for inspiration.
  –When the diaphragm contracts, the abdominal contents are pushed downward, and the ribs are lifted upward and outward.

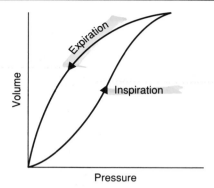

Figure 4-2. Compliance of the lung. Different curves are followed during inspiration and expiration (hysteresis).

2. External intercostals and accessory muscles
  –are not used for inspiration during normal quiet breathing.
  –are used during **exercise**.

B. Muscles of expiration
  –Expiration is **normally passive**.
  –Because the lung/chest wall system is elastic, it returns to its resting position after inspiration.
  –Expiratory muscles are used **during exercise** or when airway resistance is increased because of disease (e.g., **asthma**).

1. Abdominal muscles
  –compress the abdominal cavity, push the diaphragm up, and push air out of the lungs.

2. Internal intercostal muscles
  –pull the ribs downward and inward.

C. Compliance in the respiratory system
  –is the **slope of the pressure–volume curve**.
  –is the change in volume for a given change in pressure.

1. Compliance of the lung (Figure 4-2 and Table 4-1)
  –Inflation of the lung (inspiration) follows a different curve than deflation of the lung (expiration); this is called **hysteresis**.
  –In the middle range of pressures, the compliance is greater—the lungs are more distensible.
  –At high expanding pressures, the compliance is lower; the lungs are less distensible and the curve flattens.

Table 4-1. Causes of Increased and Decreased Lung Compliance

Causes of ↓ Lung Compliance	Causes of ↑ Lung Compliance
High expanding pressures	Emphysema (↓ elastic fibers)
↑ pulmonary venous pressure	Age
Fibrosis (deposition of collagen)	
Lack of surfactant	

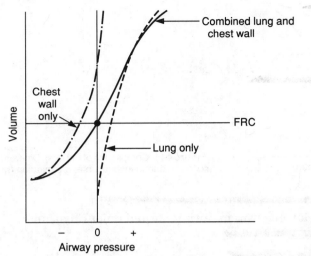

Figure 4-3. Compliance of the lung and chest wall separately and together.

2. Compliance of combined lung–chest wall system (Figure 4-3)

a. At rest (identified by the filled circle in the center of Figure 4-3), lung volume is at FRC and the pressure in the lungs is equal to atmospheric pressure. Under these equilibrium conditions, the **lungs have a tendency to collapse, which is exactly balanced by the tendency of the chest wall to spring out**.

b. As a result of these two opposing tendencies, **intrapleural pressure is subatmospheric or negative**.

c. If air is introduced into the intrapleural space (**pneumothorax**), the intrapleural pressure becomes equal to atmospheric pressure. The lung will collapse (its natural tendency) and the chest wall will spring outward (its natural tendency).

d. Figure 4-3 shows the pressure–volume relationships for the lung alone (hysteresis has been eliminated for simplicity), the chest wall alone, and the lung and chest wall together.

- The **compliance of the lung–chest wall system is less** than that of the lungs alone or the chest wall alone (the slope is flatter).
- In a patient with **emphysema,** lung compliance is increased and the tendency of the lungs to collapse is decreased. Therefore, at the original FRC, the tendency of the lungs to collapse is much less than the tendency of the chest wall to expand. The lung–chest wall system will seek a new, higher FRC so that the two forces can be balanced again; the patient develops a **barrel-shaped chest** in order to provide this new volume.

D. Surface tension of alveoli and surfactant

1. Surface tension of alveoli (Figure 4-4)

- results from the forces between molecules of liquid lining the alveoli.
- Because of their surface tension, the alveoli have a tendency to collapse.

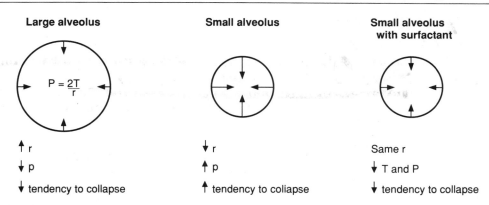

Figure 4-4. Effect of alveolar size and surfactant on the pressure that tends to collapse them.

—The pressure tending to collapse alveoli is directly proportional to surface tension and inversely proportional to alveolar radius (**Laplace's law**), as shown in the following equation:

$$P = \frac{2T}{r}$$

P	**collapsing pressure on alveolus**	**dynes/cm^2**
	or	
	pressure required to keep alveolus open	
T	**surface tension**	**dynes/cm**
r	**radius of alveolus**	**cm**

a. **Large alveoli** (big radius) have low collapsing pressures and are easy to keep open.

b. **Small alveoli** (small radius) have higher collapsing pressures and are harder to keep open.

   —In the **absence of surfactant,** the small alveoli have a tendency to collapse (**atelectasis**).

2. **Surfactant** (see Figure 4-4)

   —lines the alveoli and reduces surface tension.

   —is made by **type II alveolar cells** and consists primarily of the lipid **dipalmitoyl phosphatidyl choline**.

   —**reduces surface tension.** This reduction prevents small alveoli from collapsing, and **increases compliance**.

   —In the **fetus,** surfactant synthesis is variable. It may occur as early as gestational week 24 and is almost always present by gestational week 35.

   —**Neonatal respiratory distress syndrome** can occur in premature infants because of the lack of surfactant. The infant exhibits **atelectasis** during expiration, difficulty reinflating the lungs (as a result of **decreased compliance**), and **hypoxemia** from the V/Q mismatch.

E. **Changes in lung compliance** (see Table 4-1)

F. Airway resistance

1. Air flow

- is driven by, and is **directly proportional to, the pressure difference** between the mouth (or nose) and the alveoli.
- is **inversely proportional to airway resistance,** as shown in the equation below. Thus, the **higher the resistance, the lower the flow**.

$$Q = \Delta P / R$$

Q	air flow	ml/min or L/min
ΔP	pressure gradient	mm Hg or cm H_2O
R	airway resistance	cm H_2O/L/sec

2. Resistance of the airways

- is described by **Poiseuille's law,** as shown in the following equation:

$$R = \frac{8 \eta l}{\pi r^4}$$

R	resistance
η	viscosity of the inspired gas
l	length of the airway
r	radius of the airway

- Notice the powerful inverse relationship between resistance and the size (radius) of the airway.
- **For example,** if airway radius decreases by a factor of 2, then resistance will increase by a factor of 16 (2^4), and air flow will decrease dramatically by a factor of 16.

3. Factors that change airway resistance

a. Contraction or relaxation of bronchial smooth muscle

- changes the radius of the airways.
  - (1) **Parasympathetic stimulation, irritants, or slow-reacting substance of anaphylaxis (asthma)** constrict the airways, decrease radius, and increase resistance to air flow.
  - (2) **Sympathetic stimulation and sympathetic agonists (isoproterenol)** dilate the airways, increase radius, and decrease resistance to air flow via β_2 **receptors**.

b. Lung volume

- affects airway resistance because of the radial traction exerted on the airways by surrounding lung tissue.
- **High lung volumes** are associated with more traction and decreased airway resistance. Patients with increased airway resistance (e.g., asthma) may "learn" to breathe at higher lung volumes to offset the high resistance.
- **Low lung volumes** are associated with less traction and increased airway resistance, even to the point of airway collapse.

c. Viscosity or density of the inspired gas

- changes resistance to air flow.

–During a deep sea dive, density and resistance to air flow are increased.

–Breathing a low-density gas such as helium reduces resistance to air flow.

4. Sites of airway resistance

–The major site of airway resistance is the **medium-sized bronchi**.

–The smallest airways would *seem* to offer highest resistance, but they do not because of their parallel arrangement.

G. Breathing cycle—description of pressures and air flow (Figure 4-5)

1. At rest (before inspiration begins)

a. Alveolar pressure equals atmospheric pressure.

–Because lung pressures are always referred to atmospheric pressure, **alveolar pressure is said to be zero**.

b. Intrapleural pressure is negative.

–The opposing forces of the lungs trying to collapse and the chest wall trying to expand create a negative pressure in the intrapleural space between them.

–Intrapleural pressure can be measured by a **balloon catheter in the esophagus**.

c. Lung volume is the FRC.

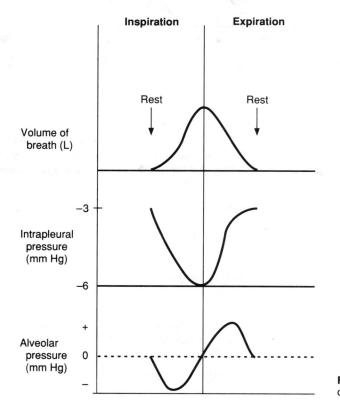

Figure 4-5. Volumes and pressures during the breathing cycle.

2. During inspiration

a. The inspiratory muscles contract and cause the volume of the thorax to increase.

—Because lung volume increases, alveolar pressure decreases below atmospheric pressure (becomes negative).

—This **pressure gradient causes air to flow into the lungs;** it will continue to flow until the pressure gradient between the atmosphere and the alveoli dissipates.

b. Intrapleural pressure becomes more negative during inspiration.

—Because lung volume increases during inspiration, the elastic recoil strength of the lungs also increases. As a result, intrapleural pressure becomes even more negative than it was at rest. Negative airway pressure also contributes to the more negative intrapleural pressure.

—Changes in intrapleural pressure during inspiration are used to measure the **dynamic compliance** of the lungs.

c. Lung volume increases by one tidal volume.

—Thus, at the peak of inspiration, lung volume is the FRC plus one tidal volume.

3. During expiration

a. Alveolar pressure becomes greater than atmospheric pressure.

—The alveolar pressure becomes greater (positive) because alveolar gas is compressed by the elastic forces of the lung.

—The pressure gradient is now reversed, and air flows out of the lungs.

b. Intrapleural pressure returns to its resting value during a normal (passive) expiration.

—During a **forced expiration,** intrapleural pressure actually becomes positive. This positive intrapleural pressure compresses the airways and makes expiration more difficult.

—In **chronic obstructive pulmonary disease (COPD),** where airway resistance is increased, patients learn to expire slowly with **"pursed lips"** to prevent the airway collapse that occurs with a forced expiration.

c. Lung volume returns to FRC before another cycle begins.

III. Gas Exchange—General Information

A. Dalton's law of partial pressures

—can be expressed by the following equation:

Partial pressure = total pressure × fractional concentration

1. In dry inspired air, the partial pressure of O_2 can be calculated as follows. Assume that total pressure is atmospheric and the fractional concentration of O_2 is 0.21.

$$P_{O_2} = 760 \text{ mm Hg} \times 0.21$$
$$= 160 \text{ mm Hg}$$

2. In humidified tracheal air at 37° C, the calculation is modified to correct for the partial pressure of H_2O = 47 mm Hg.

$$P_{total} = 760 \text{ mm Hg} - 47 \text{ mm Hg}$$
$$= 713 \text{ mm Hg}$$

$$P_{O_2} = 713 \text{ mm Hg} \times 0.21$$
$$= 150 \text{ mm Hg}$$

B. Partial pressures of O_2 and CO_2 (Table 4-2)

—About 2% of the systemic cardiac output (bronchial and coronary flow) bypasses the pulmonary circulation (**"physiologic shunt"**).

—The resulting admixture of venous blood with oxygenated arterial blood makes the P_{O_2} of arterial blood slightly less than that of alveolar air.

C. Dissolved gases

—The amount of gas dissolved in a solution (such as blood) is proportional to its partial pressure. The units of concentration for a dissolved gas are ml gas/100 ml blood.

—Use O_2 as an **example**:

$$[O_2] = P_{O_2} \times \text{solubility of } O_2 \text{ in blood}$$
$$= 100 \text{ mm Hg} \times 0.03 \text{ ml } O_2/L/\text{mm Hg}$$
$$= 0.3 \text{ ml } O_2/100 \text{ ml blood}$$

D. Diffusion of gases such as O_2 and CO_2

—The diffusion rates of O_2 and CO_2 **depend on the partial pressure differences** across the membrane and the area available for diffusion.

—**For example,** the diffusion of O_2 from alveolar air into pulmonary capillary blood depends on the partial pressure difference for O_2 across pulmonary capillaries. Normally, capillary blood equilibrates with alveolar gas; when the partial pressures of O_2 become equal (see Table 4-2), then there is no more net diffusion of O_2.

Table 4-2. Partial Pressures of O_2 and CO_2 (mm Hg)

Gas	Dry Inspired Air	Humidified Tracheal Air	Alveolar Air	Systemic Arterial Blood	Mixed Venous Blood
P_{O_2}	160	150	100	100*	40
		Addition of H_2O decreases P_{O_2}	O_2 has diffused from alveolar air into pulmonary capillary blood, decreasing the P_{O_2} of alveolar air	Blood has equilibrated with alveolar air (is "arterialized")	O_2 has diffused from arterial blood into tissues, decreasing the P_{O_2} of venous blood
P_{CO_2}	0	0	40	40	46
			CO_2 has been added from pulmonary capillary blood into alveolar air	Blood has equilibrated with alveolar air	CO_2 has diffused from the tissues into venous blood, increasing the P_{CO_2} of venous blood

*Actually, slightly < 100 mm Hg because of "physiologic shunts."

Table 4-3. Perfusion-limited and Diffusion-limited Gas Exchange

Perfusion-limited Gases	Diffusion-limited Gases
O_2 (normal conditions)	O_2 (emphysema, fibrosis, exercise)
CO_2	CO
N_2O	

E. Perfusion-limited and diffusion-limited gas exchange (Table 4-3)

1. **Perfusion-limited exchange**
   - When the gas **equilibrates** early along the length of the pulmonary capillary, the partial pressure of arterial blood becomes equal to the partial pressure of alveolar air.
   - Thus, diffusion of the gas can be increased if blood flow increases.

2. **Diffusion-limited exchange**
   - When the gas **does not equilibrate** fully by the time blood reaches the end of the pulmonary capillary, the partial pressure in arterial blood is less than that of alveolar air.
   - Diffusion of O_2 from alveolar air to pulmonary capillary blood is usually perfusion-limited but becomes diffusion-limited in disease.

   a. In **fibrosis,** diffusion of O_2 is decreased because of thickening of the alveolar membrane and increased diffusion distance.

   b. In **emphysema,** the diffusion of O_2 is decreased because the surface area for diffusion is decreased.

IV. Oxygen Transport

- O_2 is carried in blood in two forms: either dissolved in solution or (most importantly) bound to hemoglobin.
- Hemoglobin, at its normal concentration, increases the O_2-carrying capacity of blood seventyfold.

A. Hemoglobin

1. **Characteristics—globular protein of four subunits**
   - Each subunit contains a **heme moiety,** which is iron-containing porphyrin.
   - The iron is in the ferrous state (**Fe^{2+}**), which binds O_2.
   - If iron is in the ferric state (**Fe^{3+}**), it is **methemoglobin,** which does not bind O_2.
   - Each subunit has a polypeptide chain. Two subunits have α chains and two subunits have β chains; thus, normal adult hemoglobin is called $\alpha_2 \beta_2$.

2. **Fetal hemoglobin**
   - In **fetal hemoglobin, the β chains are replaced by γ chains**.
   - The O_2 affinity of fetal hemoglobin is higher than the O_2 affinity of adult hemoglobin because 2,3-diphosphoglycerate (DPG) is bound less avidly. Because the O_2 affinity of fetal hemoglobin is higher than the O_2 affinity of adult hemoglobin, O_2 movement from mother to fetus is facilitated (IV C 2 b).

3. **O_2 capacity**
   - is determined by the hemoglobin concentration in blood.
   - limits the amount of O_2 that can be carried in blood.

4. O$_2$ content

   —is the amount of O$_2$ carried in blood.
   —depends upon the hemoglobin concentration, the P$_{O_2}$, and the P$_{50}$ of hemoglobin.

B. Hemoglobin–O$_2$ dissociation curve (Figure 4-6)

   1. Hemoglobin combines rapidly and reversibly with O$_2$ to form **oxyhemoglobin**.

   2. The hemoglobin–O$_2$ dissociation curve is a plot of P$_{O_2}$ versus percent saturation of hemoglobin.

   a. At a P$_{O_2}$ of 100 mm Hg (arterial blood)
   —Hemoglobin is almost 100% saturated; O$_2$ is bound to all four heme groups on all hemoglobin molecules.

   b. At a P$_{O_2}$ of 40 mm Hg (mixed venous blood)
   —Hemoglobin is 75% saturated, which means that, on average, three of the four heme groups on each hemoglobin molecule have O$_2$ bound.

   c. At a P$_{O_2}$ of 25 mm Hg
   —Hemoglobin saturation is only 50%.
   —The P$_{O_2}$ at 50% saturation is the **P$_{50}$**. Fifty percent saturation means that, on average, two of the four heme groups of each hemoglobin molecule have O$_2$ bound.

   3. The **sigmoid shape** of the curve is the result of a change in affinity of hemoglobin as each successive O$_2$ molecule binds to a heme site; binding of the first O$_2$ molecule increases the affinity for the second O$_2$ molecule, and so forth. **The affinity for the fourth O$_2$ molecule is the highest.**
   —This change in affinity facilitates the loading of hemoglobin with O$_2$ in the lungs (flat portion of curve) and the unloading of O$_2$ at the tissues (steep portion of curve).

   a. In the lungs
   —Alveolar gas has a P$_{O_2}$ of 100 mm Hg.

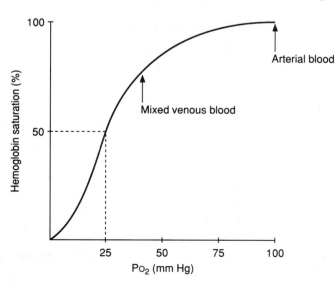

Figure 4-6. Hemoglobin–O$_2$ dissociation curve.

−Pulmonary capillary blood is "arterialized" by the diffusion of O_2 from alveolar gas into blood, so that its Po_2 also becomes 100 mm Hg.

−The very high affinity of hemoglobin for O_2 in this range facilitates the diffusion process by tightly binding O_2 and thus maintaining the partial pressure gradient (which drives diffusion of gas).

−**Because the curve is almost flat in the Po_2 range from 60 mm Hg to 100 mm Hg,** humans can tolerate changes in atmospheric pressure (and Po_2) without compromising the O_2-carrying capacity of hemoglobin.

b. In the peripheral tissues

−O_2 diffuses from arterial blood to the cells.

−The gradient for O_2 diffusion is maintained because the cells consume O_2 for aerobic metabolism, keeping their Po_2 low.

−The lower affinity of hemoglobin for O_2 in this steeper portion of the curve facilitates the unloading of O_2 to the tissues.

C. Changes in the hemoglobin–O_2 dissociation curve (Figure 4-7)

1. Shifts to the right

−occur when the **affinity of hemoglobin for O_2 is decreased**.

−The **P_{50} is increased,** and unloading of O_2 from arterial blood to the tissues is facilitated.

−For any level of Po_2, the percent saturation of hemoglobin is decreased.

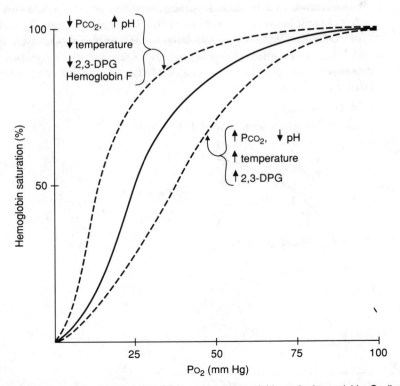

Figure 4-7. Effects of Pco_2; pH; temperature; 2,3-DPG; and fetal hemoglobin on the hemoglobin–O_2 dissociation curve.

a. Increases in P_{CO_2} or decreases in pH

- –shift the curve to the right, decreasing the affinity of hemoglobin for O_2 and facilitating the unloading of O_2 in the tissues (**Bohr effect**).
- –**For example,** during exercise the tissues produce more CO_2, which lowers the tissue pH and, by the Bohr effect, stimulates O_2 delivery to the exercising muscle.

b. Increases in temperature (e.g., during exercise)

- –shift the curve to the right.
- –The shift to the right decreases the affinity of hemoglobin for O_2 and facilitates the delivery of O_2 to the tissues during this period of high demand.

c. Increases in 2,3-DPG concentration

- –shift the curve to the right by binding to the β chains of deoxyhemoglobin and decreasing the affinity for O_2.
- –The **adaptation to chronic hypoxemia** (e.g., living at high altitude) includes increased synthesis of 2,3-DPG, which binds to hemoglobin to facilitate unloading of O_2 in the tissues.

2. Shifts to the left

–occur when the **affinity of hemoglobin for O_2 is increased.**
–The **P_{50} is decreased, and unloading of O_2 from arterial blood into the tissues is more difficult.**
–For any level of P_{O_2}, the percent saturation of hemoglobin is increased.

a. Causes of a shift to the left

- –are the mirror image of those causing a shift to the right: **decreased P_{CO_2}, increased pH, decreased temperature, and decreased 2,3-DPG concentration.**

b. Fetal hemoglobin (hemoglobin F)

- –does not bind 2,3-DPG as strongly as does adult hemoglobin. Therefore, the affinity of fetal hemoglobin for O_2 is increased, the P_{50} is decreased, and the curve is **shifted to the left.**

c. Carbon monoxide poisoning (Figure 4-8)

- –Carbon monoxide competes for O_2 binding sites on hemoglobin. The **affinity of hemoglobin for carbon monoxide is much higher** than its affinity for O_2.
- –The hemoglobin–O_2 dissociation curve is **shifted to the left,** indicating a higher affinity of the remaining sites for O_2 and, therefore, impaired ability to unload O_2 in the tissues.
- –The **total O_2 content of blood is reduced** because the O_2-carrying capacity of hemoglobin is compromised by carbon monoxide.

D. Causes of hypoxemia (Table 4-4)

V. CO_2 Transport

A. Forms of CO_2

–CO_2 is produced in the tissues and carried to the lungs in the venous blood in three forms:

1. Dissolved CO_2 (small amount)

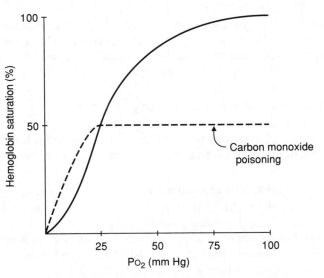

Figure 4-8. Effect of carbon monoxide on the hemoglobin–O_2 dissociation curve.

2. Carbaminohemoglobin (small amount)

3. HCO_3^- (from hydration of CO_2 in red blood cells) is the **major form (90%)**.

B. Transport of CO_2 as HCO_3^- (Figure 4-9)

   1. **CO_2 is generated in the tissues** and diffuses freely into venous plasma and then into the red blood cells.

   2. In the red blood cells, CO_2 combines with H_2O to form H_2CO_3, a reaction that is catalyzed by **carbonic anhydrase**. H_2CO_3 dissociates into H^+ and HCO_3^-.

   3. The HCO_3^- leaves the red blood cells in exchange for Cl^- (**chloride shift**) and is transported to the lungs in the plasma. HCO_3^- is the major form in which CO_2 is transported to the lungs.

   4. The H^+ is buffered inside the red blood cells by deoxyhemoglobin. Because **deoxyhemoglobin is a better buffer for H^+ than is oxyhemoglobin**, it is advantageous that hemoglobin has been deoxygenated by the time blood reaches the venous end of the capillaries where CO_2 is being added.

Table 4-4. Mechanisms and Examples of Hypoxemia

Mechanism	Example
↓ alveolar P_{O_2}	Living at high altitude
Hypoventilation	Neuromuscular disease Sedatives Chronic obstructive pulmonary disease
V/Q mismatch	Fibrosis Pulmonary embolism Pulmonary edema
Venous admixture	Right-to-left cardiac shunt
↓ O_2-carrying capacity	Anemia Carbon monoxide poisoning

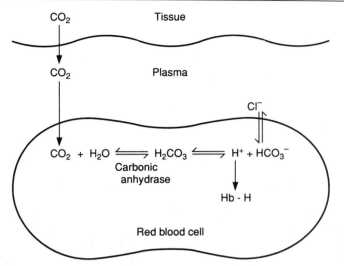

Figure 4-9. Transport of CO_2 from the tissues to the lungs. Hb-H is H^+ buffered by hemoglobin.

5. In the lungs, all of the above reactions occur in reverse. HCO_3^- enters red blood cells in exchange for Cl^-. HCO_3^- recombines with H^+ to form H_2CO_3, which decomposes into CO_2 and H_2O. CO_2, originally generated in the tissues and carried to the lungs as HCO_3^-, is expired.

VI. Pulmonary Circulation

A. Pressures and cardiac output in the pulmonary circulation

1. Pressures

–are **much lower than in the systemic circulation**.

2. Resistance

–is **also much lower** in the pulmonary circulation than in the systemic circulation.

3. Cardiac output

–equals cardiac output through the systemic circulation.

–The low pressure of the pulmonary circulation is sufficient to pump these high levels of cardiac output because pulmonary resistance is proportionately low.

B. Distribution of pulmonary blood flow

–The distribution of blood flow in the lungs is uneven and is explained by the **effects of gravity**.

–When a person is **supine,** blood flow is nearly uniform throughout the lung.

–When a person is **standing,** blood flow is **lowest at the apex of the lung (zone 1)** and **highest at the base of the lung (zone 3)** [Figure 4-10].

1. Zone 1—blood flow is lowest.

–Alveolar pressure > arterial pressure > venous pressure.

–The high alveolar pressure collapses the capillaries and causes reduced flow. This situation can occur when arterial blood pressure is decreased as a result of **hemorrhage** or when alveolar pressure is increased because of **positive pressure ventilation**.

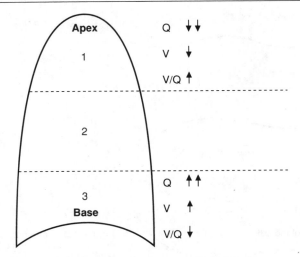

Figure 4-10. Variation of ventilation and blood flow (perfusion) in different regions of the lung.

2. Zone 2—blood flow is medium.
  —Arterial pressure > alveolar pressure > venous pressure.
  —Moving down the lung, arterial pressure progressively increases because of the effect of gravity on hydrostatic pressure.
  —Arterial pressure is now greater than alveolar pressure, and blood flow is driven by the difference between arterial pressure and alveolar pressure.

3. Zone 3—blood flow is highest.
  —Arterial pressure > venous pressure > alveolar pressure.
  —Moving down toward the base of the lung, arterial pressure is highest because of gravitational effects, and venous pressure finally increases to the point where it exceeds alveolar pressure.
  —Now blood flow is driven by the difference between arterial and venous pressures.

C. Regulation of pulmonary blood flow—hypoxic vasoconstriction
  —**Hypoxia causes local vasoconstriction.** This response is the **opposite of that in the systemic circulation** where hypoxia causes vasodilation.
  —Physiologically, this effect is important because local vasoconstriction diverts blood away from poorly ventilated, hypoxic regions of the lung (e.g., **following bronchial obstruction**) and toward well-ventilated regions.
  —**Fetal pulmonary vascular resistance** is very high because of generalized hypoxic vasoconstriction; as a result, blood flow through the fetal lungs is low. With the first breath, the alveoli of the neonate become better oxygenated, pulmonary vascular resistance decreases, and pulmonary blood flow becomes equal to cardiac output (as occurs in the adult).

D. Shunts

1. Right-to-left shunts
  —normally occur to a small extent because 2% of the cardiac output bypasses the lungs.
  —may be 50% of cardiac output in certain congenital abnormalities.
  —always result in a **decrease in arterial P_{O_2}** because of the admixture of venous blood with arterial blood.

–The degree of right-to-left shunt can be estimated by having a patient breathe 100% O_2 and measuring the degree of equilibration of arterial P_{O_2}.

2. Left-to-right shunts

–are much **less common** than right-to-left shunts.

–are usually caused by congenital abnormalities (e.g., **patent ductus arteriosus**) or traumatic injury.

–**do not result in a decrease in arterial P_{O_2}.** Instead, P_{O_2} will be elevated on the right side of the heart because there has been admixture of arterial blood with venous blood.

VII. Ventilation/Perfusion Defects

A. Ventilation/perfusion (V/Q) ratio

–is the **ratio of alveolar ventilation (V) to pulmonary blood flow (Q)**. Matching ventilation and perfusion is important to achieve ideal exchange of O_2 and CO_2.

1. Normal V/Q ratio

–If frequency, tidal volume, and pulmonary cardiac output are normal, the V/Q ratio is about 0.8. This results in values of 100 mm Hg for arterial P_{O_2} and 40 mm Hg for arterial P_{CO_2}.

2. V/Q ratio in airway obstruction

–If the airways are completely blocked (e.g., by a piece of steak caught in the trachea), then ventilation is zero. If blood flow is normal, then **V/Q is zero**.

–There is **no gas exchange** in a lung that is perfused but not ventilated. The P_{O_2} **and P_{CO_2} of pulmonary capillary blood** (and, therefore, of systemic arterial blood) will **approach that of mixed venous blood.**

3. V/Q ratio in blood flow obstruction

–If blood flow to a lung is completely blocked (e.g., by an embolism occluding a pulmonary artery), then blood flow to that lung is zero. If ventilation is normal, then **V/Q is infinite**.

–There is **no gas exchange** in a lung that is ventilated but not perfused. The P_{O_2} **and P_{CO_2} of alveolar gas will approach that of inspired air**.

4. V/Q ratios in different parts of the lung (Table 4-5; see Figure 4-10)

–Both ventilation and blood flow are distributed nonuniformly in the normal lung.

a. Blood flow is lowest at the apex and highest at the base.

b. Ventilation is also lowest at the apex and highest at the base, but the regional differences for ventilation are not as great as for perfusion.

c. Therefore, the **V/Q ratio is highest (greater than 1.0) at the apex of the lung** and **lowest (less than 0.8) at the base of the lung**.

Table 4-5. V/Q Characteristics of Different Areas of Lung

Area of Lung	Blood Flow	Ventilation	V/Q	Regional Arterial P_{O_2}	Regional Arterial P_{CO_2}
Apex	Lowest	Lower	Highest	↑	↓
Base	Highest	Higher	Lowest	↓↓	↑

    d. As a result of the regional differences in V/Q ratio, there are corresponding differences in alveolar and pulmonary capillary P_{O_2} and P_{CO_2}. **Regional differences in P_{O_2} are greater than for P_{CO_2}.**

       (1) At the apex (highest V/Q), P_{O_2} is highest and P_{CO_2} is lowest.

       (2) At the base (lowest V/Q), P_{O_2} is lowest and P_{CO_2} is highest.

       (3) When the **lungs are diseased,** V/Q inequalities are more pronounced than in the normal lung; consequently, there can be severe **hypoxemia** and modest **hypercapnia.**

VIII. Control of Breathing

—Sensory information (P_{CO_2}, lung stretch, irritant, muscle spindles, tendons, and joints) is coordinated in the **brainstem.**

—The output of the brainstem controls the respiratory muscles and the breathing cycle.

A. Central control of breathing (brainstem and cortex)

1. Medullary respiratory center

—is located in the **reticular formation.**

a. Dorsal respiratory group

—is primarily responsible for **inspiration** and **generates the basic rhythm for breathing.**

—**Input** to the dorsal respiratory group comes from vagus and glossopharyngeal nerves. The vagus nerve relays information from peripheral chemoreceptors and mechanoreceptors in the lung. The glossopharyngeal nerve relays information from peripheral chemoreceptors.

—**Output** from the dorsal respiratory group travels, via the phrenic nerve, to the diaphragm.

b. Ventral respiratory group

—is primarily responsible for expiration.

—is not active during normal, quiet breathing since expiration is passive.

—is activated, for example, during exercise when expiration becomes an active process.

2. Apneustic center

—is located in the **lower pons.**

—**stimulates inspiration,** producing a deeper and more prolonged inspiratory gasp (apneusis).

3. Pneumotaxic center

—is located in the **upper pons.**

—**inhibits inspiration** and, therefore, **regulates inspiratory volume and respiratory rate.**

4. Cortex

—Breathing can be under voluntary control; therefore, a person can voluntarily hyperventilate or hypoventilate.

—Hypoventilation (breath-holding) is limited by the increase in P_{CO_2} and decrease in P_{O_2}. A previous period of hyperventilation extends the period of breath-holding.

B. Chemoreceptors for CO_2, H^+, and O_2 (Table 4-6)

1. Central chemoreceptors in the medulla

—are sensitive to the **pH** of cerebrospinal fluid (CSF).

—H^+ does not cross the blood–brain barrier as well as CO_2 does.

a. CO_2 diffuses from arterial blood into the CSF because CO_2 is lipid-soluble and readily crosses the blood–brain barrier.

b. In the CSF, CO_2 combines with H_2O to produce H^+ and HCO_3^-. **The resulting H^+ acts directly on the central chemoreceptors.**

c. Thus, **increases in P_{CO_2} and $[H^+]$ stimulate breathing,** and decreases in P_{CO_2} and $[H^+]$ inhibit breathing.

d. The resulting hyperventilation or hypoventilation then returns the arterial P_{CO_2} toward normal.

2. Peripheral chemoreceptors in the carotid and aortic bodies

—The carotid bodies are located at the bifurcation of the common carotid arteries.

—The aortic bodies are located above and below the aortic arch.

a. Decreases in arterial P_{O_2}

—stimulate the peripheral chemoreceptors and **increase breathing rate**.

—P_{O_2} must fall to low levels (< **60 mm Hg**) before breathing is stimulated. Below 60 mm Hg, breathing rate is very sensitive to P_{O_2}.

b. Increases in arterial P_{CO_2}

—stimulate peripheral chemoreceptors and **increase breathing rate**.

—potentiate the stimulation of breathing caused by hypoxemia.

—The response of the peripheral chemoreceptors to CO_2 is not as important as the response of the central chemoreceptors to CO_2 (H^+).

c. Increases in arterial $[H^+]$

—stimulate carotid body peripheral chemoreceptors directly, independent of changes in P_{CO_2}.

—Thus, in metabolic acidosis where arterial $[H^+]$ is increased (pH decreased), breathing rate is increased.

C. Other types of receptors for control of breathing

1. Lung stretch receptors

—are located in the smooth muscle of the airways.

—When they are stimulated by distention of the lungs, they produce a reflex decrease in breathing frequency (**Hering-Breuer reflex**).

Table 4-6. Comparison of Central and Peripheral Chemoreceptors

	Location	Stimuli that Increase Breathing Rate
Central chemoreceptors	Medulla	↓ pH (↑ P_{CO_2})
Peripheral chemoreceptors	Carotid and aortic bodies	↓ P_{O_2} (if < 60 mm Hg) ↑ P_{CO_2} ↓ pH

2. Irritant receptors

–are located between airway epithelial cells.

–are stimulated by noxious substances.

3. J receptors (juxtacapillary)

–are located in the alveolar walls close to the capillaries.

–Engorgement of pulmonary capillaries, such as might occur with **left heart failure,** stimulates the J receptors, which then cause rapid, shallow breathing.

4. Joint and muscle receptors

–are activated during movement of the limbs.

–are involved in the early stimulation of breathing during **exercise**.

IX. Integrated Responses of the Respiratory System

A. Exercise (Table 4-7)

1. There is an **increase in ventilatory rate** that **matches the increase in O$_2$ consumption and CO$_2$ production** by the body.

2. The *mean* **values for arterial P$_{O_2}$ and P$_{CO_2}$ do not change** during exercise. Arterial pH does not change during moderate exercise, although strenuous exercise may cause a decrease in pH because of **lactic acidosis**.

3. On the other hand, **venous P$_{CO_2}$ increases** during exercise because the excess CO$_2$ produced by the exercising muscle is carried to the lungs in venous blood.

4. The stimulus for the increased ventilation rate during exercise is unknown. Joint and muscle receptors are activated during movement and cause an increase in breathing rate at the beginning of exercise. Oscillations in arterial P$_{CO_2}$ may stimulate the peripheral chemoreceptors to increase the breathing rate.

5. **Pulmonary blood flow increases** because cardiac output increases during exercise. As a result, more pulmonary capillaries are perfused and there is more gas exchange. Because there is **more even distribution of V/Q ratios** throughout the lung during exercise than when at rest, there is a **decrease in the physiologic dead space**.

Table 4-7. Summary of Respiratory Responses to Exercise

Parameter	Response
O$_2$ consumption	↑
CO$_2$ production	↑
Ventilation rate	↑ (matches O$_2$ consumption/CO$_2$ production)
Arterial P$_{O_2}$ and P$_{CO_2}$	No change
Arterial pH	No change in moderate exercise
	↓ in strenuous exercise (lactic acidosis)
Venous P$_{CO_2}$	↑
Pulmonary blood flow (cardiac output)	↑
V/Q ratios	More evenly distributed in lung

Table 4-8. Summary of Adaptation to High Altitude

Parameter	Response
Alveolar P_{O_2}	↓ (resulting from ↓ barometric pressure)
Arterial P_{O_2}	↓ (hypoxemia)
Ventilation rate	↑ (hyperventilation)
Arterial pH	↑ (respiratory alkalosis)
Hemoglobin concentration	↑
2,3-DPG concentration	↑
Hemoglobin–O_2 curve	Shift to right; ↓ affinity
Pulmonary vascular resistance	↑

B. Adaptation to high altitude (Table 4-8)

1. **Alveolar P_{O_2} is decreased at high altitude** because the barometric pressure is decreased. As a result, arterial P_{O_2} is also decreased (**hypoxemia**).

2. Hypoxemia stimulates the peripheral chemoreceptors and increases the ventilation rate (**hyperventilation**). This hyperventilation produces **respiratory alkalosis**. The respiratory alkalosis can be treated by administering **acetazolamide**. → ↑ Renal excretion of Bicarb.

3. Hypoxemia also stimulates renal production of **erythropoietin,** which increases the production of red blood cells. As a result, there is **increased hemoglobin concentration** and increased O_2-carrying capacity of blood.

4. **2,3-DPG concentrations are increased,** producing a shift to the right of the hemoglobin–O_2 dissociation curve. The resulting decreased affinity of hemoglobin for O_2 facilitates unloading of O_2 in the tissues.

5. **Pulmonary vasoconstriction** is another result of the hypoxemia. Consequently, there is an increase in pulmonary artery pressure, increased work of the right heart against the higher resistance, and hypertrophy of the right heart.

6. cellular changes (↑ mitochondria)

Review Test

Directions: Each of the numbered items or incomplete statements in this section is followed by answers or by completions of the statement. Select the **one** lettered answer or completion that is **best** in each case.

1. Which of the following lung volumes or capacities CANNOT be measured by spirometry?

(A) Tidal volume
(B) Inspiratory reserve volume
(C) Expiratory reserve volume
(D) Inspiratory capacity
(E) Functional residual capacity

2. An infant born prematurely in gestational week 25 is found to have neonatal respiratory distress syndrome. Which of the following would NOT be expected in this infant?

(A) Cyanosis
(B) Collapse of small alveoli
(C) Decreased lung compliance
(D) Dyspnea
(E) Lecithin:sphingomyelin ratio of greater than 2:1 in amniotic fluid prior to birth

3. In which vascular bed does hypoxia cause vasoconstriction?

(A) Coronary
(B) Pulmonary
(C) Cerebral
(D) Muscle
(E) Skin

Questions 4 and 5

A 12-year-old boy suffers a severe asthmatic attack with wheezing. He experiences rapid breathing and becomes cyanotic. His arterial P_{O_2} is 60 mm Hg and P_{CO_2} is 30 mm Hg.

4. Which of the following is most likely to be true about this patient?

(A) FEV_1/FVC is increased
(B) V/Q ratio is increased in affected areas of his lungs
(C) His arterial P_{CO_2} is increased above normal because of inadequate gas exchange
(D) His arterial P_{CO_2} is decreased below normal because his hypoxemia is causing him to hyperventilate
(E) His residual volume is decreased

5. To treat this patient, the physician should administer

(A) an α-adrenergic antagonist
(B) a β_1-adrenergic antagonist
(C) a β_2-adrenergic agonist
(D) a muscarinic agonist
(E) a nicotinic agonist

6. Which of the following is true during inspiration?

(A) Intrapleural pressure is positive
(B) The volume in the lungs is less than FRC
(C) Alveolar pressure equals atmospheric pressure
(D) Alveolar pressure is higher than atmospheric pressure
(E) Intrapleural pressure is more negative than it is during expiration

7. Residing at high altitude causes all of the following EXCEPT

(A) hyperventilation
(B) hypoxemia
(C) increased 2,3-DPG concentrations
(D) shift to the left of the hemoglobin–O_2 dissociation curve
(E) pulmonary vasoconstriction

8. When a person is standing, the distribution of blood flow in the lungs is

(A) equal at the apex and at the base
(B) highest at the apex due to the effects of gravity on arterial pressure
(C) highest at the base because the difference between arterial and venous pressure is greatest there
(D) lowest at the base because alveolar pressure is greater than arterial pressure there

9. All of the following cause hypoxemia
EXCEPT

(A) anemia
(B) pulmonary fibrosis
(C) left-to-right cardiac shunt
(D) right-to-left cardiac shunt
(E) residing at high altitude

10. Which of the following is illustrated in the graph showing volume versus pressure in the lung/chest wall system?

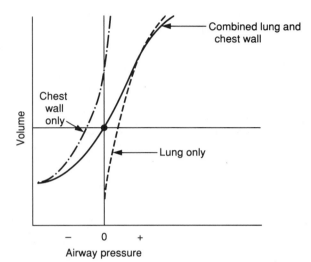

(A) The slope of each of the curves is resistance
(B) The compliance of the lungs alone is less than the compliance of the combined lung plus chest wall
(C) The compliance of the chest wall alone is less than the compliance of the lungs plus chest wall
(D) When airway pressure is zero (atmospheric), the volume in the combined lung plus chest wall is the FRC
(E) When airway pressure is zero (atmospheric), intrapleural pressure is zero

11. In the transport of CO_2 from the tissues to the lungs, all of the following occur in venous blood EXCEPT

(A) conversion of CO_2 and H_2O to H^+ and HCO_3^- in the red blood cells
(B) buffering of H^+ by deoxyhemoglobin
(C) shift of HCO_3^- out of red blood cells into plasma in exchange for Cl^-
(D) binding of CO_2 to hemoglobin
(E) alkalinization of red blood cells

12. Which of the following is the site of highest airway resistance?

(A) Mouth
(B) Largest bronchi
(C) Medium sized bronchi
(D) Smallest bronchi
(E) Alveoli

13. If blood flow to the left lung is completely blocked by an embolism in its pulmonary artery, which of the following will occur?

(A) V/Q ratio in the left lung will be zero
(B) Systemic arterial P_{O_2} will be elevated
(C) V/Q ratio in the left lung will be lower than in the right lung
(D) Alveolar P_{O_2} in the left lung will be approximately equal to the P_{O_2} in inspired air

Questions 14 and 15

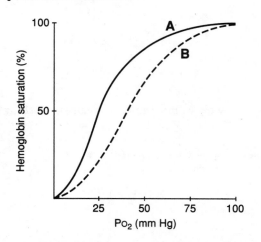

14. In the hemoglobin–O_2 dissociation curves shown above, the shift from curve A to curve B could be caused by

(A) increased pH
(B) decreased 2,3-DPG concentration
(C) strenuous exercise
(D) fetal hemoglobin
(E) carbon monoxide

15. The shift from curve A to curve B is associated with

(A) increased P_{50}
(B) increased affinity of hemoglobin for O_2
(C) impaired ability to unload O_2 in the tissues
(D) increased O_2-carrying capacity of hemoglobin
(E) decreased O_2-carrying capacity of hemoglobin

16. The pH of venous blood is only slightly more acid than the pH of arterial blood because

(A) CO_2 is a base
(B) there is no carbonic anhydrase in venous blood
(C) the H^+ generated from CO_2 and H_2O is buffered by HCO_3^- in venous blood
(D) the H^+ generated from CO_2 and H_2O is buffered by deoxyhemoglobin in venous blood
(E) oxyhemoglobin is a better buffer for H^+ than is deoxyhemoglobin

17. Compared with the systemic circulation, the pulmonary circulation has a

(A) higher flow
(B) lower resistance
(C) higher arterial pressure
(D) higher capillary pressure
(E) higher cardiac output

18. Compared with the apex of the lung, the base of the lung has

(A) a higher pulmonary capillary P_{O_2}
(B) a higher pulmonary capillary P_{CO_2}
(C) a higher V/Q ratio
(D) the same V/Q ratio

19. Hypoxemia produces hyperventilation by a direct effect on the

(A) phrenic nerve
(B) J receptors
(C) lung stretch receptors
(D) medullary chemoreceptors
(E) carotid and aortic body chemoreceptors

20. Which of the following changes occurs during strenuous exercise?

(A) Ventilation rate increases to the same extent as O_2 consumption increases
(B) Systemic arterial P_{O_2} decreases to about 70 mm Hg
(C) Systemic arterial P_{CO_2} increases to about 60 mm Hg
(D) Systemic venous P_{CO_2} decreases to about 20 mm Hg
(E) Pulmonary blood flow decreases

21. If an area of the lung is not ventilated because of bronchial obstruction, the pulmonary capillary blood serving that area will have a P_{O_2} that is

(A) equal to atmospheric P_{O_2}
(B) equal to mixed venous P_{O_2}
(C) equal to normal systemic arterial P_{O_2}
(D) higher than inspired P_{O_2}
(E) less than mixed venous P_{O_2}

Directions: Each group of items in this section consists of lettered options followed by a set of numbered items. For each item, select the **one** lettered option that is most closely associated with it. Each lettered option may be selected once, more than once, or not at all.

Questions 22 and 23

Match each numbered description below to the correct lung volume or capacity.

(A) Tidal volume
(B) Vital capacity
(C) Expiratory reserve volume
(D) Residual volume
(E) Functional residual capacity

22. Volume remaining in the lungs after expiring a tidal volume

23. Volume remaining in the lungs after a maximal expiration

Answers and Explanations

1–E. Residual volume cannot be measured by spirometry. Therefore, any lung volume or capacity that includes the residual volume cannot be measured by spirometry. These are the residual volume, the functional residual capacity, and the total lung capacity.

2–E. Neonatal respiratory distress syndrome is caused by lack of adequate surfactant in the immature lung. Surfactant appears variably, from the twenty-fourth to the thirty-fifth gestational week. In the absence of surfactant, the surface tension of the small alveoli is too high. When the pressure on the small alveoli is too high (P = 2T/r), the small alveoli collapse into larger alveoli. There is decreased gas exchange with the larger, collapsed alveoli causing V/Q mismatch, hypoxemia, and cyanosis. The lack of surfactant also decreases lung compliance, making it harder to inflate the lungs and increasing the work of breathing. Generally, lecithin:sphingomyelin ratios greater than 2:1 signify mature levels of surfactant.

3–B. Pulmonary blood flow is controlled locally by the P_{O_2} of alveolar air. *Hypoxia causes pulmonary vasoconstriction* and thereby shunts blood away from unventilated areas of the lung where it would be wasted. In the coronary circulation, hypoxemia causes vasodilation. Cerebral, muscle, and skin circulations are not controlled directly by P_{O_2}.

4–D. The patient's arterial P_{CO_2} is lower than the normal value of 40 mm Hg because hypoxemia has stimulated peripheral chemoreceptors to increase his breathing rate; hyperventilation results in blowing off extra CO_2 and respiratory alkalosis. In an obstructive disease such as asthma, both FEV_1 and FVC are decreased, with the larger decrease in FEV_1. Therefore, the ratio of FEV_1/FVC is decreased. Poor ventilation of affected areas decreases V/Q ratios and causes hypoxemia. The patient's residual volume is increased because breathing at a higher lung volume partially offsets the increased resistance of the airways.

5–C. The cause of the airway obstruction in asthma is bronchiolar constriction. β_2-adrenergic stimulation (β_2-agonists) produces relaxation of the bronchioles.

6–E. During inspiration, intrapleural pressure becomes *more negative* than it is at rest or during expiration (when it returns to its less negative resting value). Air flows into the lungs during inspiration because the alveolar pressure becomes lower (due to the efforts of the diaphragm) than atmospheric pressure; if it were not lower than atmospheric pressure, air would not flow inward. The volume in the lungs during inspiration is the FRC *plus* one tidal volume.

7–D. At high altitudes, the P_{O_2} of alveolar air is decreased because barometric pressure is decreased. As a result, there is hypoxemia, which causes hyperventilation by an effect on peripheral chemoreceptors. 2,3-DPG levels increase adaptively; 2,3-DPG binds to hemoglobin and causes a **shift to the right** of the hemoglobin–O_2 dissociation curve to improve unloading of O_2 in the tissues. The pulmonary vasculature constricts in response to hypoxia.

8–C. The distribution of blood flow in the lungs is affected by gravitational effects on arterial hydrostatic pressure. Thus, flow is highest at the base where the arterial hydrostatic pressure is greatest; there, the difference between arterial and venous pressure is greatest (which is what drives the blood flow).

9–C. High altitude causes hypoxemia because atmospheric and alveolar P_{O_2} are decreased when barometric pressure is decreased. Right-to-left shunt causes admixture of venous blood with oxygenated arterial blood, thus lowering the P_{O_2} of arterial blood. In anemia, the hemoglobin concentration and, therefore, the O_2-carrying capacity of blood is decreased. In fibrosis, there is impaired diffusion of O_2 across the alveolar membranes. Left-to-right heart shunt results in admixture of arterial blood with venous blood and does not decrease arterial P_{O_2}. Instead, P_{O_2} will be elevated on the right side of the heart.

10–D. By convention, when airway pressure is equal to atmospheric pressure, it is designated as zero pressure. Under these equilibrium conditions, there is no airflow because there is no pressure gradient between the atmosphere and the alveoli. The volume in the lungs at equilibrium is the FRC. The slope of each curve is compliance, not resistance; the steeper the slope, the greater the volume change for a given pressure change, or the greater compliance. The compliance of the lungs alone or the chest wall alone is greater than the combined lung/chest wall system (the slopes are steeper, which means higher compliance). When airway pressure is zero (equilibrium conditions), intrapleural pressure is negative because of the opposing tendencies of the chest wall to spring out and the tendency of the lungs to collapse.

11–E. CO_2 generated in the tissues is hydrated to form H^+ and HCO_3^- in red blood cells. The H^+ is buffered inside the red cell by hemoglobin, which *acidifies* the red blood cells. HCO_3^- leaves the red blood cell in exchange for Cl^- and is carried to the lungs in the plasma. A small amount of CO_2 binds directly to hemoglobin (carbaminohemoglobin).

12–C. The medium-sized bronchi actually constitute the site of highest resistance along the bronchial tree. While the small radii of the alveoli might predict that they would have the highest resistance, this is not the case because of their parallel arrangement. In fact, early changes in resistance in the small airways may be "silent" and go undetected because their overall contribution to resistance is small.

13–D. Alveolar P_{O_2} in the left lung will equal the P_{O_2} in inspired air. Because there is no blood flow to the left lung, there can be no gas exchange between the alveolar air and the pulmonary capillary blood. Consequently, O_2 is not delivered to the capillary blood. *V/Q ratio in the left lung will be infinite* (not zero or lower than in the normal right lung) since Q (the denominator) is zero. Systemic arterial P_{O_2} will, of course, be decreased since one lung has no gas exchange.

14–C. Strenuous exercise increases temperature and decreases pH of skeletal muscle; both effects would cause a shift to the right of the hemoglobin–O_2 dissociation curve, making it easier to unload O_2 in the tissues to meet the high demand of the exercising muscle. 2,3-DPG binds to the β chains of adult hemoglobin and reduces its affinity for O_2, shifting the curve to the right. In fetal hemoglobin, the β chains are replaced by γ chains, which do not bind 2,3-DPG, so the curve is shifted to the left. Because carbon monoxide increases the affinity of the remaining binding sites for O_2, the curve is shifted to the left.

15–A. A shift to the right of the hemoglobin–O_2 dissociation curve represents decreased affinity of hemoglobin for O_2. At any given P_{O_2}, the percent saturation is decreased, the P_{50} is increased (read the P_{O_2} from the graph at 50% hemoglobin saturation), and unloading of O_2 in the tissues is facilitated. The O_2-carrying capacity of hemoglobin is determined by hemoglobin concentration and is unaffected by the shift from curve A to curve B.

16–D. In venous blood, CO_2 combines with H_2O and produces the weak acid H_2CO_3, catalyzed by carbonic anhydrase. The resulting H^+ is buffered by deoxyhemoglobin, which is such an effective buffer for H^+ (meaning that the pK is within 1.0 unit of the pH of blood) that the pH of venous blood is only slightly more acid than the pH of arterial blood. In fact, oxyhemoglobin is a *less effective buffer* than deoxyhemoglobin.

17–B. Flow (or cardiac output) through the systemic and pulmonary circulations is nearly equal; pulmonary flow is slightly less than the systemic flow since about 2% of the systemic cardiac output bypasses the lungs. The pulmonary circulation is characterized by both *lower pressure and lower resistance* than the systemic circulation, so flows through the two circulations are about equal (flow = pressure/resistance).

18–B. Ventilation and perfusion of the lung are not distributed uniformly. Both are lowest at the apex and highest at the base. However, the differences for ventilation are not as great as for perfusion, so V/Q ratios are higher at the apex and lower at the base. As a result, gas exchange is more efficient at the apex and less efficient at the base. Therefore, blood at the apex will have a higher Po_2 and lower Pco_2 since it is better equilibrated with alveolar air.

19–E. Hypoxemia stimulates breathing by a direct effect on the peripheral chemoreceptors in the carotid and aortic bodies. Central (medullary) chemoreceptors are stimulated by CO_2 (H^+). The J receptors and lung stretch receptors are not chemoreceptors. The phrenic nerve innervates the diaphragm, and its activity is determined by the output of the brainstem.

20–A. During exercise, the ventilation rate increases to match the increased O_2 consumption and CO_2 production. This matching is accomplished *without a change in mean arterial Po_2 or Pco_2*. Venous Pco_2 does increase because extra CO_2 is being produced by the exercising muscle. Because this CO_2 will be blown off by the hyperventilating lungs, it does not raise the arterial Pco_2. Pulmonary blood flow (cardiac output) increases many-fold during strenuous exercise.

21–B. If an area of lung is not ventilated, there can be no gas exchange in that region. The pulmonary capillary blood serving that region will not equilibrate with alveolar Po_2, but will have a Po_2 equal to that of mixed venous blood.

22–E. During normal breathing, the volume inspired and then expired is a tidal volume. The volume remaining in the lungs after expiration of a tidal volume is the functional residual capacity.

23–D. During a forced maximal expiration, the volume expired is a tidal volume *plus* the expiratory reserve volume. The volume remaining in the lungs is the residual volume.

5
Renal and Acid–Base Physiology

I. Body Fluids

—Total body water (TBW) is approximately **60% of body weight**.
—The percentage of TBW is **highest in newborns and adult males** and **lowest in adult females and in adults with a large amount of adipose tissue**. (Obese)

A. Distribution of water (Figure 5-1 and Table 5-1)

1. Intracellular fluid (ICF)

—is **2/3 of TBW**.
—The major cations of ICF are K^+ and Mg^{2+}.
—The major anions of ICF are **protein and organic phosphates** (adenosine triphosphate [ATP], adenosine diphosphate [ADP], and adenosine monophosphate [AMP]).

2. Extracellular fluid (ECF)

—is **1/3 of TBW**.
—is comprised of interstitial fluid and plasma.
—The major cation of ECF is Na^+.
—The major anions of ECF are Cl^- and HCO_3^-

a. Plasma comprises **1/4 of the ECF**. Thus, it is **1/12 of TBW** (1/4 × 1/3).

—The major **plasma proteins** are albumin and globulins.

b. Interstitial fluid comprises **3/4 of the ECF**. Thus, it is **1/4 of TBW** (3/4 × 1/3).

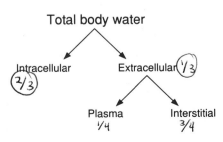

Figure 5-1. Summary of body fluid compartments.

135

Table 5-1. Summary of Body Fluid Compartments

Body Fluid Compartment	Fraction of TBW*	Markers Used to Measure Volume	Major Cations	Major Anions
TBW	1.0	Tritiated H_2O D_2O		
ECF	1/3	Sulfate Inulin Mannitol	Na^+	Cl^- HCO_3^-
Plasma	1/12 (1/4 of ECF)	Radioiodinated serum albumin (RISA) I^{131}-albumin Evans blue	Na^+	Cl^- HCO_3^- Plasma protein
Interstitial	1/4 (3/4 of ECF)	ECF − plasma volume (indirect)	Na^+	Cl^- HCO_3^-
ICF	2/3	TBW − ECF (indirect)	K^+	Organic phosphates Protein

*Total body water (TBW) is approximately 60% of total body weight or 42 liters in a 70-kg male.
ECF = extracellular fluid; ICF = intracellular fluid.

—The composition of interstitial fluid is the same as that of plasma except that it has **little protein**. Thus, interstitial fluid is an **ultrafiltrate of plasma**.

B. Measuring the volumes of the fluid compartments (see Table 5-1)

1. Dilution method

a. **A known amount of a substance is given** whose volume of distribution is the body fluid compartment of interest.

—**Examples are:**

(1) Mannitol is a marker substance for ECF because it is a large molecule that cannot cross cell membranes and is excluded from the ICF.

(2) Evans blue is a marker for plasma volume because it is a dye that binds to serum albumin and is confined to the plasma compartment.

b. The substance is allowed to **equilibrate**.

c. The **concentration of the substance is measured** in plasma, and the **volume of distribution is calculated** as follows:

$$\text{Volume} = \frac{\text{amount}}{\text{concentration}}$$

Volume	volume of distribution or volume of the body fluid compartment	L
Amount	amount of substance injected	mg
Concentration	concentration in plasma	mg/L

d. Sample calculation

—A patient is injected with 500 mg of mannitol. After a 2-hour equilibration period, the concentration of mannitol in plasma is 3.2 mg/100 ml. During the equilibration period, 10% of the injected mannitol is excreted in urine. What is the ECF volume?

$$\text{Volume} = \frac{\text{amount}}{\text{concentration}}$$

$$= \frac{\text{amount injected} - \text{amount excreted}}{\text{concentration}}$$

$$= \frac{500 \text{ mg} - 50 \text{ mg}}{3.2 \text{ mg/100 ml}}$$

$$= 14.1 \text{ L}$$

2. Substances used for major fluid compartments (see Table 5-1)

a. TBW

—Tritiated water and D_2O

b. ECF

—Sulfate, inulin, and mannitol

c. Plasma

—Radioiodinated serum albumin (RISA) and Evans blue

d. Interstitial

—Measured indirectly (ECF − plasma)

e. ICF

—Measured indirectly (TBW − ECF)

C. Shifts of water between compartments

1. Basic principles

a. **Water shifts between ECF and ICF** so that the osmolarities of the two compartments become equal.

b. The **osmolarity of ICF and ECF** are assumed to be **equal** after a brief period of equilibration.

c. It can be assumed that solutes such as NaCl and mannitol do not cross cell membranes and are confined to ECF.

2. Examples of shifts of water between compartments (Figure 5-2 and Table 5-2)

a. Infusion of isotonic NaCl—addition of isotonic fluid

—is also called **isosmotic volume expansion.**

(1) ECF volume increases, but there will be **no change in osmolarity** of ECF or ICF. Because osmolarity is unchanged, there will be no shift of water between ECF and ICF.

(2) Plasma protein concentration and hematocrit decrease because the addition of fluid to the ECF dilutes the protein and red blood cells. Because ECF osmolarity is unchanged, the red blood cells will not shrink or swell.

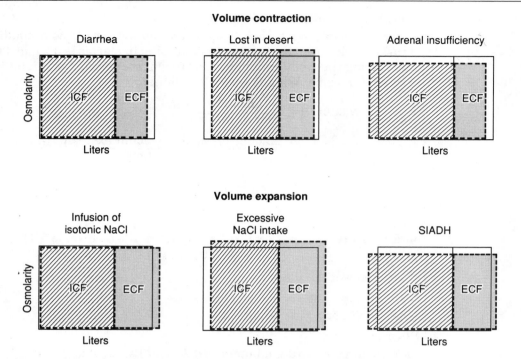

Figure 5-2. Shifts of water between body fluid compartments. Normal ECF and ICF volume and osmolarity are indicated by the *solid lines*. Changes in volume and osmolarity in response to various situations are indicated by the *dashed lines*. ECF = extracellular fluid; ICF = intracellular fluid.

       (3) Arterial blood pressure increases because ECF volume increases.

  b. Diarrhea—loss of isotonic fluid

    —is also called **isosmotic volume contraction**.

    (1) ECF volume decreases, but there will be **no change in osmolarity** of ECF or ICF. Because osmolarity is unchanged, there will be no shift of water between ECF and ICF.

    (2) Plasma protein concentration and hematocrit increase because the loss of ECF concentrates the protein and red blood cells. Because ECF osmolarity is unchanged, the red blood cells will not shrink or swell.

    (3) Arterial blood pressure decreases because ECF volume decreases.

  c. Excessive NaCl intake—addition of NaCl

    —is also called **hyperosmotic volume expansion**.

    (1) Osmolarity of ECF increases because there has been addition of osmoles to the ECF.

    (2) Water shifts from ICF to ECF. As a result of this shift, **ICF osmolarity increases** until it is equal to ECF osmolarity.

    (3) As a result of the shift of water out of the cells, the **volume of the ECF increases** (volume expansion) and the **volume of the ICF decreases**.

Table 5-2. Changes in Volume and Osmolarity of Body Fluids

Type	Key Examples	ECF Volume	ICF Volume	ECF Osmolarity	Hct, Serum [Na$^+$]
Isosmotic volume expansion	Isotonic NaCl infusion	↑	No change	No change	↓ Hct − [Na$^+$]
Isosmotic volume contraction	Diarrhea	↓	No change	No change	↑ Hct − [Na$^+$]
Hyperosmotic volume expansion	High NaCl intake	↑	↓	↑	↓ Hct ↑ [Na$^+$]
Hyperosmotic volume contraction	Sweating Fever Diabetes insipidus	↓	↓	↑	− Hct ↑ [Na$^+$]
Hyposmotic volume expansion	SIADH	↑	↑	↓	− Hct ↓ [Na$^+$]
Hyposmotic volume contraction	Adrenal insufficiency	↓	↑	↓	↑ Hct ↓ [Na$^+$]

  d. Sweating in a hot desert—loss of water
   −is also called **hyperosmotic volume contraction**.
   (1) Osmolarity of ECF increases because sweat is hyposmotic (relatively more water than salt is lost during sweating).
   (2) ECF volume decreases because of the loss of volume in the sweat. Water shifts out of the ICF. As a result of the shift, **ICF osmolarity increases** until it is equal to ECF osmolarity.
   (3) As a result of the shift of water out of the cells, **ICF volume decreases**.
   (4) Protein concentration increases because of the loss of ECF volume. Although **hematocrit** might also be expected to increase, in fact it is **unchanged** because water shifts out of the red blood cells, decreasing their volume and offsetting the concentrating effect of ECF volume contraction.

  e. Syndrome of inappropriate antidiuretic hormone (SIADH)—gain of water
   −is also called **hyposmotic volume expansion**.
   (1) Osmolarity of ECF decreases because of retention of excess water.
   (2) ECF volume increases because of the water retention. Water shifts into the cells; as a result of this shift, **ICF osmolarity decreases** until it is equal to ECF osmolarity, and **ICF volume increases**.
   (3) Protein concentration decreases because of the increase in ECF volume. Although **hematocrit** might also be expected to decrease, in fact it will be **unchanged** because water shifts into the red blood cells, increasing their volume and offsetting the diluting effect of ECF volume expansion.

  f. Adrenocortical insufficiency—loss of NaCl − aldosterone
   −is also called **hyposmotic volume contraction**.

(1) **Osmolarity of ECF decreases** because the kidneys lose more NaCl than water as a result of the lack of aldosterone in adrenocortical insufficiency.

(2) **ECF volume decreases.** Water shifts into the cells; as a result of this shift, **ICF osmolarity decreases** until it equals ECF osmolarity, and **ICF volume increases**.

(3) **Protein concentration increases** because of the decrease in ECF volume. **Hematocrit increases** because of the decreased ECF volume and the swelling of red blood cells caused by water entry.

II. Renal Clearance, Renal Plasma Flow, and Glomerular Filtration Rate

A. Clearance equation

–indicates the volume of plasma cleared of a substance per unit time.

–The units of clearance are **ml/min** or **ml/24 hr**.

$$C = \frac{UV}{P}$$

C	clearance	ml/min or ml/24 hr
U	urine concentration	mg/ml
V	urine volume/time	ml/min
P	plasma concentration	mg/ml

–**Example of clearance calculation:** If the serum $[Na^+]$ is 140 mEq/L, the urine $[Na^+]$ is 700 mEq/L, and the urine flow rate is 1 ml/min, what is the clearance of Na^+?

$$C_{Na^+} = \frac{[U]_{Na^+} \times V}{[P]_{Na^+}}$$

$$= \frac{700 \text{ mEq/L} \times 1 \text{ ml/min}}{140 \text{ mEq/L}}$$

$$= 5 \text{ ml/min}$$

B. Renal blood flow (RBF)

–is **25% of the cardiac output**.

–is directly proportional to the pressure difference between the renal artery and the renal vein, and is inversely proportional to the resistance of the renal vasculature.

1. Autoregulation of RBF

–is accomplished by **changing renal vascular resistance** as arterial pressure changes, thus maintaining a constant blood flow.

–RBF remains constant over the range of arterial pressures from 100–200 mm Hg (**autoregulation**).

2. Measurement of renal plasma flow (RPF)—clearance of para-aminohippuric acid (PAH)

–PAH is both **filtered and secreted** by the renal tubules.

–The clearance of PAH is used to measure RPF.

—Clearance of PAH measures **effective RPF** since it underestimates true RPF by 10%. Clearance does not include plasma flow to regions of the kidney that do not filter and secrete PAH.

$$RPF = C_{PAH} = \frac{[U]_{PAH}\, V}{[P]_{PAH}}$$

RPF	renal plasma flow	ml/min or ml/24 hr
C_{PAH}	clearance of PAH	ml/min or ml/24 hr
$[U]_{PAH}$	urine concentration	mg/ml
V	urine flow rate	ml/min or ml/24 hr
$[P]_{PAH}$	plasma concentration	mg/ml

3. Measurement of RBF

$$RBF = \frac{RPF}{1 - \text{hematocrit}}$$

C. Glomerular filtration rate (GFR)

1. Measurement of GFR—clearance of inulin

—Inulin is **filtered but not reabsorbed or secreted** by the renal tubules.

—The clearance of inulin is used to measure GFR.

$$GFR = \frac{[U]_{inulin}\, V}{[P]_{inulin}}$$

GFR	glomerular filtration rate	ml/min or ml/24 hr
$[U]_{inulin}$	urine concentration	mg/ml
V	urine flow rate	ml/min or ml/24 hr
$[P]_{inulin}$	plasma concentration	mg/ml

—**Sample calculation of GFR:** Inulin is infused in a patient to achieve a steady-state plasma concentration of 1 mg/ml. A urine sample after 1 hour has a volume of 60 ml and an inulin concentration of 120 mg/ml. What is the patient's GFR?

$$
\begin{aligned}
GFR &= \frac{[U]_{inulin}\, V}{[P]_{inulin}} \\
&= \frac{120\ \text{mg/ml} \times 60\ \text{ml/hr}}{1\ \text{mg/ml}} \\
&= \frac{120\ \text{mg/ml} \times 1\ \text{ml/min}}{1\ \text{mg/ml}} \\
&= 120\ \text{ml/min}
\end{aligned}
$$

2. Estimates of GFR with blood urea nitrogen (BUN) and plasma [creatinine]

—Both BUN and plasma [creatinine] increase when GFR decreases.

—**GFR decreases with age,** although plasma [creatinine] remains constant because of decreased muscle mass.

3. Filtration fraction

—is the fraction of the RPF that is filtered across the glomerular capillaries.

–is **normally about 0.20**. Thus, 20% of the RPF is filtered. The remaining 80% leaves the glomerular capillaries by the efferent arterioles and becomes the peritubular capillary circulation.

$$\text{Filtration fraction} = \frac{\text{GFR}}{\text{RPF}}$$

4. **Determining GFR—Starling forces** (Figure 5-3)
   –The driving force for glomerular filtration is the **net ultrafiltration pressure** across the glomerular capillaries.
   –**Filtration is always favored.**
   –GFR can be expressed by the Starling equation:

$$\textbf{GFR} = \textbf{K}_f \left[(\textbf{P}_{GC} - \textbf{P}_{BS}) - (\pi_{GC} - \pi_{BS}) \right]$$

   a. **GFR** is filtration across the glomerular capillaries.

   b. **K_f is the filtration coefficient** of the glomerular capillaries.
      –The glomerular barrier consists of capillary endothelium, basement membrane, and the filtration slits of the podocytes.
      –**Anionic glycoproteins line the filtration barrier** and restrict the filtration of negatively charged plasma proteins. In **glomerular disease,** the anionic charges on the barrier may be removed and protein-uria results.

   c. **P_{GC} is glomerular capillary hydrostatic pressure,** which is constant along the length of the capillary.
      –It is **increased by dilation of the afferent arteriole or constriction of the efferent arteriole.** Increases in P_{GC} cause increases in net ultrafiltration pressure and increases in GFR.

   d. **P_{BS} is Bowman's space hydrostatic pressure** and is analogous to P_i in systemic capillaries.
      –It is **increased by constriction or blockage of the ureters.** Increases in P_{BS} cause a decrease in net ultrafiltration pressure and a decrease in GFR.

   e. **π_{GC} is glomerular capillary oncotic pressure.**

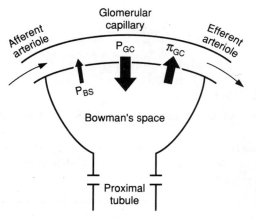

Figure 5-3. Starling forces across the glomerular capillaries. *Heavy arrows* indicate the driving forces across the glomerular capillary wall.

−Normally, it **increases along the length of the glomerular capillary** because filtration of water increases the protein concentration of glomerular capillary blood.

−It is **increased by increases in protein concentration.** Increases in π_{GC} cause a decrease in net ultrafiltration pressure and a decrease in GFR.

f. π_{BS} **is Bowman's space oncotic pressure** and is usually zero because the small amounts of filtered albumin are subsequently reabsorbed.

5. **Sample calculation of ultrafiltration pressure using the Starling equation**

−At the arteriolar end of a glomerular capillary, P_{GC} is 45 mm Hg, P_{BS} is 10 mm Hg, and π_{GC} is 27 mm Hg. What is the net ultrafiltration pressure?

$$\text{Net pressure} = (P_{GC} - P_{BS}) - \pi_{GC}$$
$$\text{Net pressure} = (45 \text{ mm Hg} - 10 \text{ mm Hg}) - 27 \text{ mm Hg}$$
$$= + 8 \text{ mm Hg (favoring filtration)}$$

6. **Changes in Starling forces—effect on GFR and filtration fraction** (Table 5-3)

III. Reabsorption and Secretion

A. **Calculation of reabsorption and secretion rates**

1. **How to calculate**

−Reabsorption or secretion rate is the difference between the amount filtered across the glomerular capillaries and the amount excreted in urine.

−If filtered load is greater than excretion rate, then **net reabsorption** of the substance has occurred. If filtered load is less than excretion rate, then **net secretion** of the substance has occurred.

Filtered load	= **GFR** × **[plasma]**
Excretion rate	= **V** × **[urine]**
Reabsorption rate	= **filtered load − excretion rate**
Secretion rate	= **excretion rate − filtered load**

Table 5-3. Effect of Changes in Starling Forces on GFR, RPF, and Filtration Fraction

	Effect on GFR	Effect on RPF	Effect on Filtration Fraction
Constriction of afferent arteriole	↓ (caused by ↓ P_{GC})	↓	No change
Constriction of efferent arteriole	↑ (caused by ↑ P_{GC})	↓	↑ (↑ GFR/↓ RPF)
Increased plasma [protein]	↓ (caused by ↑ π_{GC})	No change	↓ (↓ GFR/unchanged RPF)
Ureteral stone	↓ (caused by ↑ P_{BS})	No change	↓ (↓ GFR/unchanged RPF)

2. Example of calculation of reabsorption/secretion rate

—A woman with diabetes mellitus, which is untreated, has a GFR of 120 ml/min, a plasma glucose concentration of 400 mg/dl, a urine glucose concentration of 2500 mg/dl, and a urine flow rate of 4 ml/min. What is her reabsorption rate of glucose?

$$
\begin{aligned}
\textbf{Filtered load} \ &= \ \text{GFR} \times \text{plasma [glucose]} \\
&= \ 120 \ \text{ml/min} \times 400 \ \text{mg/dl} = 480 \ \text{mg/min} \\
\textbf{Excretion} \ &= \ \text{V} \times \text{urine [glucose]} \\
&= \ 4 \ \text{ml/min} \times 2500 \ \text{mg/dl} = 100 \ \text{mg/min} \\
\textbf{Reabsorption} \ &= \ 480 \ \text{mg/min} - 100 \ \text{mg/min} \\
&= \ \textbf{380 mg/min}
\end{aligned}
$$

B. Transport maximum (T_m) curve for glucose—a reabsorbed substance (Figure 5-4)

1. Filtered load of glucose

—increases in direct proportion to the plasma glucose concentration (filtered load of glucose = GFR × $P_{glucose}$).

2. Reabsorption of glucose

a. Na–glucose cotransport in the **proximal tubule** reabsorbs glucose from tubular fluid into the blood. There are a limited number of Na–glucose carriers.

b. At plasma glucose concentrations less than 300 mg/dl, all of the filtered glucose can be reabsorbed since there are many carriers available; in this range, the line for reabsorption is the same as that for filtration.

c. At plasma glucose concentrations of 350 mg/dl or greater, the carriers are saturated. Therefore, increases in plasma concentration do not result in increased rates of reabsorption. The point at which the carriers are saturated is the **transport maximum (T_m)**.

3. Excretion of glucose

a. At plasma concentrations below 300 mg/dl, all of the filtered glucose is

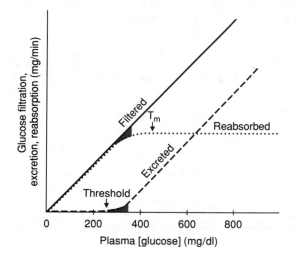

Figure 5-4. Glucose titration curve. Glucose filtration, excretion, and reabsorption are shown as a function of plasma [glucose]. *Shaded area* indicates the "splay."

reabsorbed and excretion is zero. **Threshold** (the plasma concentration at which glucose first appears in the urine) is approximately 300 mg/dl.

    b. At plasma concentrations above 350 mg/dl (T_m), reabsorption is saturated. Therefore, as plasma concentration increases, the additional filtered glucose cannot be reabsorbed and is excreted in the urine.

  4. Splay

    –is the region of the glucose curves **between threshold and T_m**.

    –occurs approximately between plasma glucose concentrations of 300 mg/dl and 350 mg/dl.

    –represents spilling of glucose in urine before saturation (T_m) is fully achieved.

    –is explained by heterogeneity of nephrons and the relatively low affinity of the glucose carriers.

C. T_m curve for PAH—a secreted substance (Figure 5-5)

  1. Filtered load of PAH

    –As with glucose, the filtered load of PAH increases in direct proportion to the plasma PAH concentration.

  2. Secretion of PAH

    a. Secretion of PAH occurs from peritubular capillary blood into tubular fluid (urine) via carriers in the **proximal tubule**.

    b. At low plasma concentrations of PAH, the secretion rate increases as the plasma concentration increases.

    c. Once the carriers are saturated, further increases in plasma PAH concentration do not cause further increase in secretion rate (**T_m**).

  3. Excretion of PAH

    a. Excretion of PAH is the **sum of filtration** across the glomerular capillaries **plus secretion** from peritubular capillary blood into urine.

    b. The curve for excretion is steepest at low plasma PAH concentrations (below T_m). Once T_m is exceeded and all of the carriers for secretion are saturated, the curve flattens and becomes parallel to the curve for filtration.

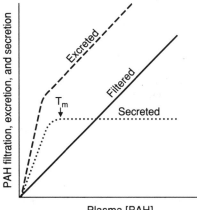

Plasma [PAH]

Figure 5-5. PAH titration curve. PAH filtration, excretion, and secretion are shown as a function of plasma [PAH].

    c. RPF is measured at plasma PAH concentrations **below the T_m**.

D. Relative clearances of substances

  1. Substances with the highest clearances

    —are those that are both filtered across the glomerular capillaries and secreted from the peritubular capillaries into urine (e.g., PAH).

  2. Substances with the lowest clearances

    —are those that are either not filtered (e.g., protein) or are filtered and subsequently reabsorbed from urine into peritubular capillary blood (e.g., Na, glucose, amino acids, HCO_3^-, Cl^-).

  3. Substances whose clearance equals GFR

    —are those that are freely filtered but not reabsorbed or secreted (e.g., inulin).

  4. Relative clearances

    —PAH > K^+ (high K^+ diet) > inulin > urea > Na > glucose, amino acids, HCO_3^-

IV. NaCl Regulation

A. Single nephron terminology

—**Tubular fluid (TF) is urine** at any point along the nephron.

—Plasma (P) refers to systemic plasma.

  1. TF/P_x ratio

    —compares the concentration of a substance in tubular fluid at a point along the nephron to its concentration in systemic plasma.

    —**For example,** if $TF/P_{Na^+} = 1.0$, then the Na^+ concentration in tubular fluid is identical to the $[Na^+]$ in plasma. For freely filtered substances, $TF/P = 1.0$ in Bowman's space (before any reabsorption or secretion has taken place to modify the tubular fluid).

    a. If TF/P = 1.0, then reabsorption has been exactly proportional to the reabsorption of water.

    b. If TF/P < 1.0, then reabsorption of the substance has occurred to a greater extent than the reabsorption of water, thus making the concentration in tubular fluid fall below that in plasma.

    c. If TF/P > 1.0, *either* **reabsorption of the substance has been less than reabsorption of water** *or* there has been **secretion of the substance**.

    d. TF/P_{inulin} is used as a marker for water reabsorption along the nephron. Since inulin is freely filtered, but not reabsorbed or secreted, its concentration in tubular fluid is determined solely by how much water remains in the tubular fluid.

      —TF/P_{inulin} increases as water is reabsorbed. **For example,** if 50% of the filtered water is reabsorbed, the $TF/P_{inulin} = 2.0$.

  2. $[TF/P]_x/[TF/P]_{inulin}$ ratio

    —corrects the TF/P_x ratio for water reabsorption. This double ratio gives the **fraction of the filtered amount remaining at any point along the nephron**.

–**For example,** if $[TF/P]_{K^+}/[TF/P]_{inulin} = 0.3$ at the end of the proximal tubule, then 30% of the filtered K^+ remains in the tubular fluid and 70% must have been reabsorbed back into blood.

B. General information about Na⁺ reabsorption

–Na^+ is freely filtered across the glomerular capillaries; the $[Na^+]$ in the tubular fluid in Bowman's space equals that in plasma ($TF/P_{Na^+} = 1.0$).

–Na^+ is reabsorbed all along the nephron and very little is excreted in urine (1% of the filtered load).

C. Na⁺ reabsorption along the nephron (Figure 5-6)

1. Proximal tubule

–**reabsorbs 2/3 or 67% of the filtered Na⁺ and H₂O,** more than any other part of the nephron.

–is the site of **glomerulotubular balance**.

–Reabsorption of Na^+ and H_2O in the proximal tubule are exactly proportional—the process is **isosmotic**. Therefore, both TF/P_{Na^+} and $TF/P_{osm} = 1.0$.

a. Early proximal tubule—special features (Figure 5-7)

–reabsorbs Na^+ and H_2O with HCO_3^-, glucose, amino acids, phosphate, and lactate.

–Na^+ is reabsorbed by **cotransport** with glucose, amino acids, phosphate, and lactate. These cotransport processes account for the reabsorption of all the filtered glucose and amino acids.

–Na^+ is also reabsorbed by **countertransport** as Na^+–H^+ exchange. Na^+–H^+ exchange is linked directly to the reabsorption of filtered HCO_3^-.

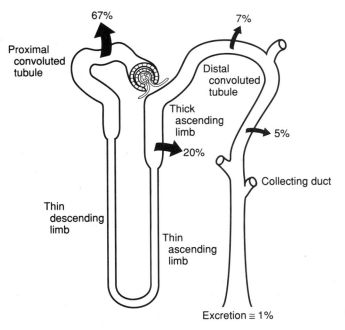

Figure 5-6. Na⁺ handling along the nephron. *Arrows* indicate reabsorption of Na⁺. *Numbers* indicate percentage of the filtered load of Na⁺ reabsorbed or excreted.

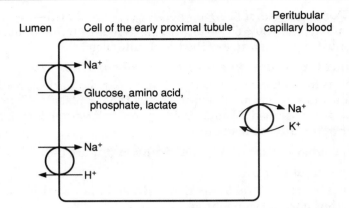

Figure 5-7. Mechanisms of Na$^+$ reabsorption in the cells of the early proximal tubule.

—**Carbonic anhydrase** inhibitors (e.g., acetazolamide) are diuretics that act in the early proximal tubule by inhibiting HCO$_3^-$ reabsorption.

b. Middle and late proximal tubules—special features
—In these segments, **Na$^+$ is reabsorbed with Cl$^-$**.
—Filtered glucose, amino acids, and HCO$_3^-$ have already been completely removed from tubular fluid by reabsorption in the early proximal tubule.

c. Glomerulotubular balance in the proximal tubule
—maintains **reabsorption of a constant fraction** (2/3) of the filtered Na$^+$ and H$_2$O.
(1) For example, if GFR increases, the filtered load of Na$^+$ increases. Glomerulotubular balance functions such that Na$^+$ reabsorption will also increase, ensuring that a constant fraction is reabsorbed.
(2) The mechanism of glomerulotubular balance resides in Starling forces, which alter Na$^+$ and H$_2$O reabsorption in the proximal tubule (Figure 5-8).
—The route of isosmotic fluid reabsorption is from the lumen, to the proximal tubule cell, to the lateral intercellular space, and then to the peritubular capillary blood.

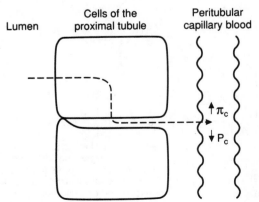

Figure 5-8. Mechanism of isosmotic reabsorption in the proximal tubule. *Dashed arrow* shows the pathway. Increases in π_c and decreases in P$_c$ cause increased rates of isosmotic reabsorption.

–**Starling forces in the peritubular capillary blood** govern how much of this isosmotic fluid will be reabsorbed. Fluid reabsorption is favored by increases in π_c of the peritubular capillary blood. Thus, increases in GFR cause the protein concentration and π_c of peritubular capillary blood to increase, which, in turn, produces an increase in fluid reabsorption—matching of filtration and reabsorption (**glomerulotubular balance**).

d. **Effects of ECF volume on proximal tubular reabsorption**

(1) **ECF volume contraction increases reabsorption.** Volume contraction increases peritubular capillary protein concentration and π_c and decreases peritubular capillary P_c. Together, these changes in Starling forces in peritubular capillary blood **cause an increase in proximal tubular reabsorption.**

(2) **ECF volume expansion decreases reabsorption.** Volume expansion decreases peritubular capillary protein concentration and π_c, and increases P_c. Together, these changes in Starling forces in peritubular capillary blood **cause a decrease in proximal tubular reabsorption.**

2. **Thick ascending limb of the loop of Henle** (Figure 5-9)

–The loop of Henle **reabsorbs 20% of the filtered Na$^+$.**

–The transport mechanism is the **Na$^+$–K$^+$–2Cl$^-$ cotransporter** in the luminal membrane of cells in the thick ascending limb.

–is the site of action of the **loop diuretics** (furosemide, ethacrynic acid, bumetanide), which inhibit the Na$^+$–K$^+$–2Cl$^-$ cotransporter.

–is **impermeable to water,** so NaCl is reabsorbed without water. As a result, tubular fluid [Na$^+$] and tubular fluid osmolarity decrease below that of plasma (TF/P$_{Na^+}$ and TF/P$_{osm}$ < 1.0). This segment, therefore, is called the **diluting segment.**

3. **Distal tubule and collecting duct**

–together **reabsorb 12% of the filtered Na$^+$.**

a. **Early distal tubule—special features**

–is called the **cortical diluting segment.**

–reabsorbs NaCl by a **Na$^+$–Cl$^-$ cotransporter.**

–is the site of action of **thiazide diuretics.**

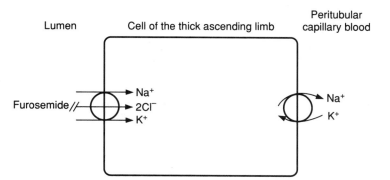

Figure 5-9. Mechanism of ion transport in the thick ascending limb of the loop of Henle.

—is **impermeable to water,** as is the thick ascending limb. Thus, the reabsorption of NaCl without water further dilutes the tubular fluid.

b. Late distal tubule and collecting duct—special features

—have two cell types.

(1) Principal cells reabsorb Na⁺ and H₂O and secrete K⁺.

—Principal cells are the site of action of the following hormones or drugs:

(a) Aldosterone increases Na⁺ reabsorption and **increases K⁺ secretion.** Like other steroid hormones, the action of aldosterone takes several hours to develop because new protein synthesis is required. About 2% of overall Na⁺ reabsorption is affected by aldosterone.

(b) ADH increases water permeability by directing the insertion of H₂O channels in the luminal membrane. In the absence of ADH, the principal cells are virtually impermeable to water.

(c) K⁺-sparing diuretics (spironolactone, triamterene, amiloride) **decrease K⁺ secretion.**

(2) Intercalated cells secrete H⁺ and reabsorb K⁺.

—**Aldosterone increases H⁺ secretion** (in addition to its actions on the principal cells).

V. K⁺ Regulation

A. Shifts of K⁺ between the ICF and ECF (Table 5-4)

—Most of the body's K⁺ is in the ICF.

—A shift of K⁺ **out of cells** causes **hyperkalemia.**

—A shift of K⁺ **into cells** causes **hypokalemia.**

B. Renal regulation of K⁺ balance (Figure 5-10)

—**K⁺ is filtered, reabsorbed, and secreted** by the nephron.

—**K⁺ balance** is achieved when urinary excretion of K⁺ exactly equals intake of K⁺ in the diet.

—K⁺ excretion can vary widely from 1%–110% of the filtered load, depending on dietary K⁺ intake, aldosterone levels, and acid–base status.

Table 5-4. Shifts of K⁺ Between ECF and ICF

Causes of Shift of K⁺ out of Cells → Hyperkalemia	Causes of Shift of K⁺ into Cells → Hypokalemia
Insulin deficiency	Insulin
β-Adrenergic antagonists	β-Agonists
Acidosis (exchange of extracellular H⁺ for intracellular K⁺)	Alkalosis (exchange of intracellular H⁺ for extracellular K⁺)
Hyperosmolarity (H₂O flows out of the cell; K⁺ diffuses out with H₂O)	Hyposmolarity (H₂O flows into the cell; K⁺ diffuses in with H₂O)
Inhibitors of Na⁺–K⁺ pump (e.g., digitalis) [when pump is blocked, K⁺ is not taken up into cells]	
Exercise	
Cell lysis	

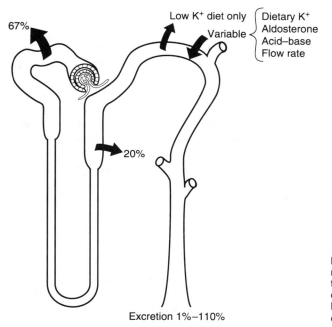

Figure 5-10. K^+ handling along the nephron. *Arrows* indicate reabsorption or secretion of K^+. *Numbers* indicate the percentages of the filtered load of K^+ reabsorbed, secreted, or excreted.

1. **Filtration**

   —occurs freely across the glomerular capillaries—TF/P_{K^+} in Bowman's space is 1.0.

2. **Proximal tubule**

   —**reabsorbs 67% of the filtered K^+** along with Na^+ and H_2O.

3. **Thick ascending limb of the loop of Henle**

   —**reabsorbs an additional 20% of the filtered K^+.**

   —Reabsorption involves the **Na^+–K^+–$2Cl^-$ cotransporter** in the luminal membrane of cells in the thick ascending limb (see Figure 5-9).

4. **Distal tubule and collecting duct** (Table 5-5)

   —either reabsorb or secrete K^+, depending on dietary K^+ intake.

   a. **Reabsorption of K^+**

      —occurs **only on a low K^+ diet** (K^+ depletion). Under these conditions, K^+ excretion can be as low as 1% of the filtered load as the kidney tries to conserve as much K^+ as possible.

Table 5-5. Effects on Distal K^+ Secretion

Causes of Increased Distal K^+ Secretion	Causes of Decreased Distal K^+ Secretion
High K^+ diet	Low K^+ diet
Hyperaldosteronism	Hypoaldosteronism
Alkalosis	Acidosis
Thiazide diuretics	K^+-sparing diuretics
Loop diuretics	
Luminal anions	

—The mechanism involves a **H$^+$–K$^+$ ATPase** in the **intercalated cells**.

b. Secretion of K$^+$

—is **variable** and accounts for the wide range for urinary K$^+$ excretion.

—depends on factors such as dietary K$^+$, aldosterone levels, acid–base status, and urine flow rate.

—occurs in the **principal cells**.

(1) Mechanism of distal K$^+$ secretion (Figure 5-11)

(a) Active transport of K$^+$ into the cell by the Na$^+$–K$^+$ pump on the basolateral membrane. This mechanism keeps the intracellular K$^+$ concentration high.

(b) Passive secretion of K$^+$ into the lumen. The magnitude of this passive secretion is **determined by the chemical and electrical driving force on K$^+$ across the luminal membrane**. Thus, maneuvers that increase the intracellular K$^+$ concentration or decrease the luminal K$^+$ concentration will increase K$^+$ secretion.

(2) Factors that change distal K$^+$ secretion (Figure 5-11)

—Distal K$^+$ secretion is increased when the electrochemical driving force for K$^+$ across the luminal membrane is increased. Secretion is decreased when the driving force is decreased.

(a) Dietary K$^+$

—A **diet high in K$^+$ increases K$^+$ secretion**, and a **diet low in K$^+$ decreases K$^+$ secretion**.

—On a **high K$^+$ diet,** intracellular K$^+$ increases so that the driving force for K$^+$ secretion also increases.

—On a **low K$^+$ diet,** K$^+$ secretion is (by the same reasoning) decreased. In fact, on a low K$^+$ diet, the intercalated cells are actually stimulated to reabsorb K$^+$.

(b) Aldosterone

—**increases K$^+$ secretion** by increasing Na$^+$ entry into the cells across the luminal membrane and increasing pumping of Na$^+$ out of the cells by the Na$^+$–K$^+$ pump.

—Stimulation of the pump simultaneously increases K$^+$ uptake into distal cells, increasing the intracellular K$^+$ concentration and the driving force for K$^+$ secretion. Thus, **hyperaldosteronism** increases K$^+$ secretion and causes hypokalemia.

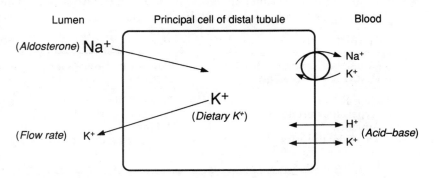

Figure 5-11. Mechanism of K$^+$ secretion in the principal cell of the distal tubule.

–**Hypoaldosteronism** decreases K^+ secretion and causes hyperkalemia.

(c) **Acid–base**

–**Acidosis decreases K^+ secretion,** and **alkalosis increases K^+ secretion**. Effectively, H^+ and K^+ exchange for each other across the cell membrane.

–In **acidosis,** because there is excess H^+ in blood, H^+ enters the distal cell in exchange for K^+ leaving the cell. As a result, intracellular K^+ concentration decreases and the driving force for K^+ secretion decreases.

–In **alkalosis,** because there is a deficit of H^+ in blood, H^+ leaves the cell in exchange for K^+ entering the cell. As a result, intracellular K^+ concentration increases and the driving force for K^+ secretion increases.

(d) **Flow rate—thiazides and loop diuretics**

–Diuretics that increase flow rate to the distal tubule (e.g., **thiazides, loop diuretics**) cause dilution of the luminal K^+ concentration, increasing the driving force for K^+ secretion. As a result of the increased K^+ secretion, these diuretics cause **hypokalemia.**

(e) **K^+-sparing diuretics—spironolactone, amiloride, triamterene**

–**inhibit K^+ secretion** and, when used alone, cause **hyperkalemia.**

–The most important use of the K^+-sparing diuretics is in combination with thiazide or loop diuretics to offset urinary K^+ losses.

–**Spironolactone** is an antagonist of aldosterone; amiloride and triamterene act directly on the distal tubule.

(f) **Luminal anions**

–The presence of excess anions in the lumen causes an increase in K^+ secretion by increasing the negativity of the lumen and increasing the electrical driving force for K^+ secretion.

VI. Renal Regulation of Urea, Phosphate, Calcium, and Magnesium

A. Urea

–**Fifty percent of the filtered urea is reabsorbed** passively in the proximal tubule.

–The distal tubule and cortical and outer medullary collecting ducts are impermeable to urea, so no urea is reabsorbed by these segments.

–**ADH increases the urea permeability of the inner medullary collecting ducts.** Urea reabsorption from inner medullary collecting ducts contributes to **urea recycling in the medulla.**

B. Phosphate

–**Eighty-five percent of the filtered phosphate is reabsorbed** in the proximal tubule by **Na^+–phosphate cotransport**. Because the distal segments of the nephron do not reabsorb phosphate, 15% of the filtered load is excreted.

–**Parathyroid hormone (PTH) inhibits phosphate reabsorption** in the proximal tubule by activation of adenylate cyclase and generation of cyclic AMP. Therefore, PTH causes **phosphaturia** and increased **urinary cyclic AMP**.

–Phosphate is a urinary buffer for H^+; excretion of $H_2PO_4^-$ is called **titratable acid**.

C. Calcium (Ca^{2+})

–**Sixty percent of the plasma Ca^{2+} is filtered** across the glomerular capillaries.

–Together, the **proximal tubule and thick ascending limb of the loop of Henle** reabsorb 90% of the filtered Ca^{2+} by passive processes that are coupled to Na^+ reabsorption.

–**Loop diuretics (e.g., furosemide, bumetanide)** produce increased urinary Ca^{2+} excretion. Because Ca^{2+} reabsorption is linked to Na^+ reabsorption in the loop of Henle, inhibition of Na^+ reabsorption with a loop diuretic also inhibits Ca^{2+} reabsorption. If volume is replaced, loop diuretics can be used to **treat hypercalcemia**.

–The **distal tubule and collecting duct,** together, reabsorb 9% of the filtered Ca^{2+} by an active process.

1. **PTH increases renal Ca^{2+} reabsorption** by activating adenylate cyclase in the distal tubule.

2. **Thiazide diuretics** produce decreased Ca^{2+} excretion because they stimulate the reabsorption of Ca^{2+} in the distal tubule. Thiazides are used to **treat hypercalciuria** by lowering urinary Ca^{2+} excretion.

D. Magnesium (Mg^{2+})

–is reabsorbed in the proximal tubule, thick ascending limb of the loop of Henle, and distal tubule.

–In the thick ascending limb, Mg^{2+} and Ca^{2+} compete for reabsorption; therefore, **hypercalcemia causes an increase in Mg^{2+} excretion** (by inhibiting Mg^{2+} reabsorption). Likewise, **hypermagnesemia causes an increase in Ca^{2+} excretion** (by inhibiting Ca^{2+} reabsorption).

VII. Concentration and Dilution of Urine

A. Regulation of plasma osmolarity

–is accomplished by **varying the amount of water excreted** by the kidney.

1. **Response to water deprivation** (Figure 5-12)

2. **Response to water intake** (Figure 5-13)

B. Concentrated urine (Figure 5-14)

–is also called **hyperosmotic urine**; urine osmolarity > blood osmolarity.

–is produced when circulating ADH levels are high (e.g., **water deprivation, hemorrhage, SIADH**).

1. **Corticopapillary osmotic gradient—high ADH**

–is the gradient of osmolarity from the cortex (300 mOsm/L) to the papilla (1200 mOsm/L), and is comprised primarily of NaCl and urea.

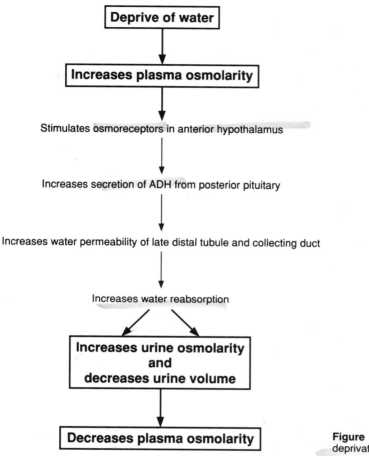

Figure 5-12. Responses to water deprivation.

—is established and maintained by the following:

a. Countercurrent multiplication in the loop of Henle requires *both* **NaCl reabsorption from the thick ascending limb of the loop of Henle** *and* **countercurrent flow**. This process is **augmented by ADH,** which stimulates NaCl reabsorption in the thick ascending limb.

b. Recycling of urea from inner medullary collecting ducts into the medullary interstitial fluid is **augmented by ADH.**

c. Vasa recta are the capillaries that supply the loop of Henle. They **maintain the corticopapillary gradient** by serving as **osmotic exchangers.** Vasa recta blood equilibrates osmotically with the interstitial fluid of the medulla and papilla.

2. Proximal tubule—high ADH

—The osmolarity of the glomerular filtrate is identical to that of plasma: 300 mOsm/L.

—Two-thirds of the filtered H_2O is reabsorbed **isosmotically** (with Na^+, Cl^-, HCO_3^-, glucose, amino acids, and so forth) in the proximal tubule.

—**$TF/P_{osm} = 1.0$** throughout the proximal tubule.

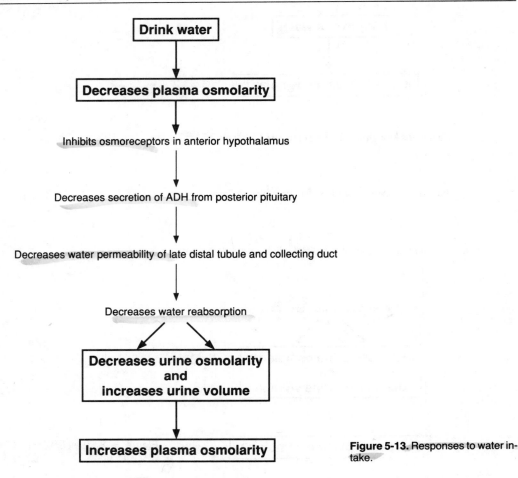

Figure 5-13. Responses to water intake.

3. Thick ascending limb of the loop of Henle—high ADH

- is called the **diluting segment**.
- reabsorbs NaCl by the **Na$^+$–K$^+$–2Cl$^-$ cotransporter**.
- is **impermeable to water;** therefore, no water is reabsorbed along with NaCl. Because NaCl is reabsorbed without H_2O, the tubular fluid becomes dilute; the fluid that leaves the thick ascending limb has an osmolarity of 100 mOsm/L and **TF/P$_{osm}$ < 1.0**.

4. Early distal tubule—high ADH

- is called the **cortical diluting segment**.
- Like the thick ascending limb, the early distal tubule reabsorbs NaCl but is **impermeable to water**. Consequently, tubular fluid is further diluted.

5. Late distal tubule—high ADH

- **Water permeability of the late distal tubule is increased by ADH.**
- Water is reabsorbed from the tubule until the osmolarity of the distal tubular fluid equals that of the surrounding interstitial fluid: 300 mOsm/L.
- **TF/P$_{osm}$ = 1.0**.

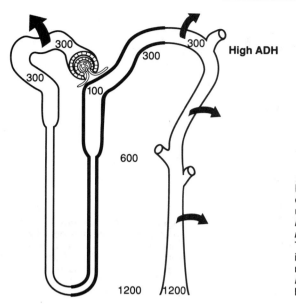

Figure 5-14. Mechanisms for producing hyperosmotic (concentrated) urine in the presence of ADH. *Numbers* indicate osmolarity. *Heavy arrows* indicate water reabsorption. The *thick outline* shows the water-impermeable segments of the nephron. (Modified from Valtin H: *Renal Function,* 2nd ed. Boston, Little, Brown & Co Inc, 1983, p 162.)

6. Collecting ducts—high ADH

- Like the late distal tubule, the **water permeability of the collecting ducts is increased by ADH**.
- As tubular fluid flows through the collecting ducts, it passes through regions of increasingly higher osmolarity (established by countercurrent multiplication and urea recycling).
- Water is reabsorbed from the tubule until the osmolarity of the tubular fluid equals that of the surrounding interstitial fluid.
- The osmolarity of the final urine equals that at the bend of the loop of Henle: 1200 mOsm/L.
- $\text{TF/P}_{osm} > 1.0$.

C. Dilute urine (Figure 5-15)

- is called **hyposmotic urine;** urine osmolarity < blood osmolarity.
- is produced when circulating levels of ADH are low (e.g., **water intake, central diabetes insipidus**) or ADH is ineffective (**nephrogenic diabetes insipidus**).

1. Corticopapillary osmotic gradient—no ADH

- is **smaller** than in the presence of ADH.

2. Proximal tubule—no ADH

- As in the presence of ADH, 2/3 of the filtered water is reabsorbed **isosmotically**.
- $\text{TF/P}_{osm} = 1.0$ throughout the proximal tubule.

3. Thick ascending limb of the loop of Henle—no ADH

- As in the presence of ADH, NaCl is reabsorbed without water and the tubular fluid becomes dilute (although not quite as dilute as in the presence of ADH).
- $\text{TF/P}_{osm} < 1.0$.

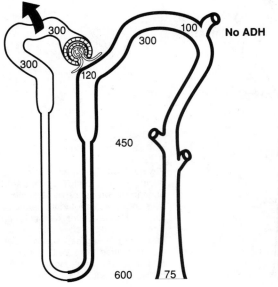

Figure 5-15. Mechanisms for producing hyposmotic (dilute) urine in the absence of ADH. *Numbers* indicate osmolarity. *Thick arrows* indicate water reabsorption. The *thick outline* shows the water-impermeable segments of the nephron. (Modified from Valtin H: *Renal Function,* 2nd ed. Boston, Little, Brown & Co Inc, 1983, p 162.)

4. Early distal tubule—no ADH

–As in the presence of ADH, NaCl is reabsorbed without water and the tubular fluid is further diluted.

5. Late distal tubule and collecting ducts—no ADH

–In the absence of ADH, the cells of the late distal tubule and collecting ducts are **impermeable to water**.
–Thus, even though the tubular fluid flows through regions of increasingly higher osmolarity, it will not equilibrate osmotically.
–The osmolarity of the **final urine will be dilute** with an osmolarity as low as 75 mOsm/L.
–**TF/P$_{osm}$ < 1.0.**

D. Use of free water clearance (C$_{H_2O}$) to estimate the ability to concentrate or dilute the urine

–Free water is produced in the diluting segments of the kidney.
–In the **absence of ADH,** this free water is excreted and there is a positive C$_{H_2O}$.
–In the **presence of ADH,** this free water is not excreted but is reabsorbed by the late distal tubule and collecting ducts, and there is a **negative C$_{H_2O}$**.

1. Calculation of C$_{H_2O}$

$$C_{H_2O} = V - C_{osm}$$

C$_{H_2O}$	free water clearance	ml/min
V	urine flow rate	ml/min
C$_{osm}$	osmolar clearance (U$_{osm}$V/P$_{osm}$)	ml/min

−**Example:** If urine flow rate is 10 ml/min, urine osmolarity is 100 mOsm/L, and plasma osmolarity is 300 mOsm/L, what is the free water clearance?

$$C_{H_2O} = V - C_{osm}$$

$$= 10 \text{ ml/min} - \frac{100 \text{ mOsm/L} \times 10 \text{ ml/min}}{300 \text{ mOsm/L}}$$

$$= 10 \text{ ml/min} - 3.33 \text{ ml/min}$$

$$= +6.7 \text{ ml/min}$$

2. Urine that is isosmotic to plasma (isosthenuric)

−C_{H_2O} is zero.

−can occur during treatment with a **loop diuretic,** which inhibits NaCl reabsorption in the thick ascending limb, and thus blocks production of the corticopapillary osmotic gradient. The urine will not be diluted (because the diluting segment is inhibited) or concentrated (because the corticopapillary gradient has been abolished).

3. Urine that is hyposmotic to plasma (low ADH)

−C_{H_2O} is positive.

−occurs with **excessive water intake** (ADH release from the pituitary is suppressed), **central diabetes insipidus** (lack of pituitary ADH), or **nephrogenic diabetes insipidus** (collecting ducts are unresponsive to ADH).

4. Urine that is hyperosmotic to plasma (high ADH)

−C_{H_2O} is **negative** (or positive free-water reabsorption [$T^c_{H_2O}$]).

−occurs in **water deprivation** (ADH release from pituitary is stimulated) or in **SIADH.**

E. Clinical disorders related to concentration and dilution of urine
(Table 5-6)

Table 5-6. Summary of Clinical Examples—ADH

	Serum ADH	Serum Osmolarity/[Na$^+$]	Urine Osmolarity	Urine Flow Rate	C_{H_2O}
1° polydipsia	↓	Decreased	Hyposmotic	High	Positive
Central diabetes insipidus	↓	Increased (because of excretion of too much H_2O)	Hyposmotic	High	Positive
Nephrogenic diabetes insipidus	↑ (because of increased plasma osmolarity)	Increased (because of excretion of too much H_2O)	Hyposmotic	High	Positive
Water deprivation	↑	High–normal	Hyperosmotic	Low	Negative
SIADH	↑ ↑	Decreased (because of reabsorption of too much H_2O)	Hyperosmotic	Low	Negative

Table 5-7. Summary of Hormones that Act on the Kidney

Hormone	Stimulus for Secretion	Time Course	Mechanism of Action	Actions on Kidneys
PTH	↓ plasma $[Ca^{2+}]$	Fast	Basolateral receptor Adenylate cyclase Cyclic AMP → urine	↓ phosphate reabsorption (proximal tubule) ↑ Ca^{2+} reabsorption (distal tubule) Stimulates 1α-hydroxylase (proximal tubule)
ADH	↑ plasma osmolarity ↓ blood volume	Fast	Basolateral V_2 receptor Adenylate cyclase Cyclic AMP (Note: V_1 receptors are on blood vessels; mechanism is Ca^{2+}– IP_3)	↑ H_2O permeability (late distal tubule and collecting duct principal cells)
Aldosterone	↓ blood volume (via renin–angiotensin II) ↑ plasma $[K^+]$	Slow	New protein synthesis	↑ Na^+ reabsorption (distal tubule principal cells) ↑ K^+ secretion (distal tubule principal cells) ↑ H^+ secretion (distal tubule intercalated cells)
Atrial natriuretic factor (ANF)	↑ atrial pressure	Fast	Guanylate cyclase Cyclic GMP	↑ GFR ↓ Na^+ reabsorption
Angiotensin II	↓ blood volume (via renin)	Fast		↑ Na^+–H^+ exchange and HCO_3^- reabsorption (proximal tubule)

VIII. Renal Hormones

—See Table 5-7 for a summary of renal hormones (see Chapter 7 for discussion of hormones).

IX. Acid–Base Balance

A. Buffers

—prevent a change in pH when H^+ ions are added or removed.

—are **most effective within 1.0 pH unit of the pK** (in the linear portion of the titration curve).

1. Extracellular buffers

a. The major extracellular buffer is **HCO_3^-**, which results from the following reaction:

$$CO_2 + H_2O \leftrightarrow H_2CO_3 \leftrightarrow H^+ + HCO_3^-$$

—**Carbonic anhydrase** catalyzes the reaction of CO_2 and H_2O; it is present in almost all cells and in the brush border of the proximal tubule.

—The **pK** of the CO_2/HCO_3^- buffer pair is 6.1.

b. Phosphate is a minor extracellular buffer.

-The **pK** of the $H_2PO_4^-$/HPO_4^{-2} buffer pair is 6.8.

-Phosphate is most important as a **urinary buffer;** excretion of H^+ as $H_2PO_4^-$ is called **titratable acid**.

2. Intracellular buffers

a. Organic phosphates (e.g., AMP, ADP, ATP, 2,3-DPG)

b. Proteins

-Imidazole and α-amino groups have a pK in the physiologic range.

-**Hemoglobin** is a major intracellular buffer.

-**Deoxyhemoglobin is a better buffer than oxyhemoglobin** in the physiologic pH range.

3. Using the Henderson-Hasselbalch equation to calculate pH

$$pH = pK + \log \frac{[A^-]}{[HA]}$$

pH	$-\log_{10} [H^+]$	**pH units**
pK	$-\log_{10}$ **equilibrium constant**	**pH units**
[A⁻]	**base form of the buffer**	**mM**
[HA]	**acid form of the buffer**	**mM**

-**Example:** The pK of the $H_2PO_4^-$/HPO_4^{-2} buffer pair is 6.8. What are the relative concentrations of $H_2PO_4^-$ and HPO_4^{-2} in a urine sample that has a pH of 4.8?

$$pH = pK + \log \frac{HPO_4^{-2}}{H_2PO_4^-}$$

$$4.8 = 6.8 + \log \frac{HPO_4^{-2}}{H_2PO_4^-}$$

$$\log \frac{HPO_4^{-2}}{H_2PO_4^-} = -2.0$$

$$\frac{HPO_4^{-2}}{H_2PO_4^-} = 0.01$$

Thus, the concentration of $H_2PO_4^-$ is 100 times that of HPO_4^{-2} in a urine sample of pH 4.8.

4. Titration curves (Figure 5-16)

-are plots of pH of a buffered solution as H^+ ions are added or removed.

-As H^+ ions are added, the HA form is produced; when H^+ ions are removed, the A^- form is produced.

-The buffer is **most effective in the linear portion** of the titration curve, when addition or removal of H^+ causes little change in pH.

-When the **pH of the solution equals the pK,** there are equal concentrations of HA and A^-.

B. Renal acid–base

1. Reabsorption of filtered HCO_3^- (Figure 5-17)

-occurs primarily in the **proximal tubule**.

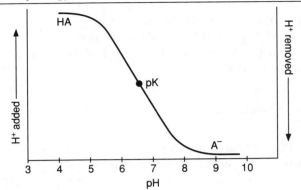

Figure 5-16. Titration curve for a weak acid (HA) and its conjugate base (A^-).

a. Key features of reabsorption of filtered HCO_3^-

(1) H^+ and HCO_3^- are produced in the proximal tubule cells when CO_2 and H_2O combine to form H_2CO_3, catalyzed by the **intracellular carbonic anhydrase**. H^+ is secreted into the lumen via the Na^+–H^+ exchange mechanism in the luminal membrane. The HCO_3^- is reabsorbed.

(2) In the lumen, the secreted H^+ combines with filtered HCO_3^- to form H_2CO_3, which dissociates to CO_2 and H_2O, catalyzed by the **brush border carbonic anhydrase**. CO_2 and H_2O diffuse into the cell to start the cycle again.

(3) There is **net reabsorption of filtered HCO_3^-**. However, this process **does not result in net secretion of H^+**.

b. Regulation of reabsorption of filtered HCO_3^-

(1) Increases in filtered load of HCO_3^- result in increased rates of HCO_3^- reabsorption. However, at very high levels of plasma HCO_3^- concentration (such as in metabolic alkalosis), the filtered load can exceed the reabsorption capacity, and HCO_3^- may be excreted in the urine.

(2) Increases in P_{CO_2} result in increased rates of HCO_3^- reabsorption because the supply of intracellular H^+ for secretion is increased. **Decreases in P_{CO_2}** result in decreased rates of reabsorption because the supply of intracellular H^+ for secretion is decreased.

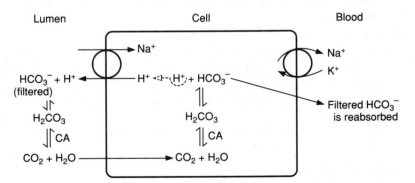

Figure 5-17. Mechanism for reabsorption of filtered HCO_3^- in the proximal tubule. *CA* = carbonic anhydrase.

−These effects of P_{CO_2} form the basis for the **renal compensation for respiratory acidosis and alkalosis**.

(3) **ECF volume contraction** results in increased HCO_3^- reabsorption (**contraction alkalosis**), and **ECF volume expansion** results in decreased HCO_3^- reabsorption.

(4) **Angiotensin II stimulates Na^+–H^+** exchange and thus increases HCO_3^- reabsorption, contributing to the contraction alkalosis that occurs with ECF volume contraction.

2. Excretion of H^+ as titratable acid ($H_2PO_4^-$) [Figure 5-18]

−is a mechanism for **excreting fixed H^+** produced from catabolism of protein and phospholipid.

a. H^+ and HCO_3^- are produced in the cell from CO_2 and H_2O. The H^+ is secreted into the lumen by an H^+–ATPase, and the HCO_3^- is reabsorbed into the blood ("new" HCO_3^-). The secreted H^+ combines with filtered HPO_4^{-2} to form $H_2PO_4^-$, which is excreted as **titratable acid**.

b. This process results in **net secretion of H^+** and **net reabsorption of newly synthesized HCO_3^-**.

c. As a result of H^+ secretion, the pH of urine becomes progressively lower. **The minimum urinary pH is 4.4.**

3. Excretion of H^+ as NH_4^+ (Figure 5-19)

−is another mechanism for excreting fixed H^+ produced from protein catabolism of protein and phospholipid.

a. NH_3 is produced in renal cells from glutamine. It diffuses down its concentration gradient into the lumen.

b. H^+ and HCO_3^- are produced in the cells from CO_2 and H_2O. The H^+ is secreted into the lumen via an H^+–ATPase and combines with NH_3 to form NH_4^+, which is excreted (**diffusion trapping**). The HCO_3^- is reabsorbed into the blood ("new" HCO_3^-).

c. The lower the pH of the tubular fluid, the greater the excretion of H^+ as NH_4^+ because diffusion of NH_3 will be favored.

d. In acidosis, there is an **adaptive increase in NH_3 synthesis,** which aids in the excretion of H^+.

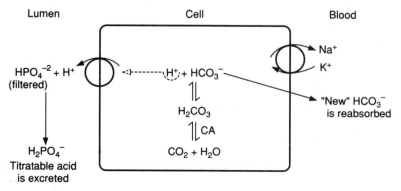

Figure 5-18. Mechanism for excretion of H+ as titratable acid. *CA* = carbonic anhydrase.

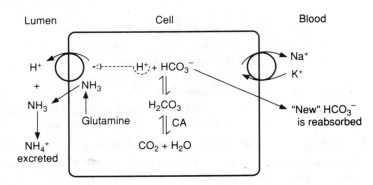

Figure 5-19. Mechanism for excretion of H^+ as NH_4^+. *CA* = carbonic anhydrase.

C. Acid–base disorders (Tables 5-8 and 5-9)

1. Metabolic acidosis

a. There is an **increase in arterial $[H^+]$** from overproduction or ingestion of fixed acid or loss of base.

b. HCO_3^- is used to buffer the added fixed acid. As a result, the arterial **$[HCO_3^-]$ decreases**.

c. Acidemia causes **hyperventilation (Kussmaul breathing), the respiratory compensation** for metabolic acidosis.

d. Renal compensation for metabolic acidosis consists of increased excretion of H^+ as titratable acid and NH_4^+, and increased reabsorption of "new" HCO_3^- (which helps replenish the HCO_3^- used in buffering the added fixed H^+).

– In **chronic metabolic acidosis,** there is an adaptive increase in NH_3 synthesis, which aids in the excretion of H^+.

e. Serum anion gap = $[Na^+] - [Cl^-] - [HCO_3^-]$. The **normal value** of the anion gap is 10–16 mEq/L and represents "other" unmeasured anions in serum.

– In metabolic acidosis, the serum $[HCO_3^-]$ decreases as it is depleted in buffering fixed acid. The concentration of another anion must increase to "replace" HCO_3^-—that anion can be Cl^- or it can be "other" unmeasured anions.

Table 5-8. Summary of Acid–Base Disorders

Disorder	$CO_2 + H_2O$ ↔	H^+ +	HCO_3^-	Respiratory Compensation	Renal Compensation
Metabolic acidosis	↓ (respiratory compensation)	↑	↓	Hyperventilation	↑ H^+ excretion ↑ "new" HCO_3^- reabsorption
Metabolic alkalosis	↑ (respiratory compensation)	↓	↑	Hypoventilation	↑ HCO_3^- excretion
Respiratory acidosis	↑	↑	↑	None	↑ H^+ excretion ↑ HCO_3^- reabsorption
Respiratory alkalosis	↓	↓	↓	None	↓ H^+ excretion ↓ HCO_3^- reabsorption

Heavy arrows indicate *primary* disturbance.

Table 5-9. Some Causes of Acid–Base Disorders

	Example	Comments
Metabolic acidosis	Ketoacidosis	Accumulation of β-OH-butyric acid and aceto-acetic acid
		↑ anion gap
	Lactic acidosis	Accumulation of lactic acid during hypoxia
		↑ anion gap
	Chronic renal failure	Failure to excrete H^+ as titratable acid and NH_4^+
		↑ anion gap
	Salicylate intoxication	Also causes respiratory alkalosis
		↑ anion gap
	Methanol/formaldehyde intoxication	Produces formic acid
		↑ anion gap
	Ethylene glycol intoxication	Produces glycolic and oxalic acids
		↑ anion gap
	Diarrhea	GI loss of HCO_3^-
		Normal anion gap
	Type 2 renal tubular acidosis (RTA)	Renal loss of HCO_3^-
		Normal anion gap
	Type 1 distal RTA	Failure to excrete titratable acid and NH_4^+; failure to acidify urine
		Normal anion gap
	Type 4 RTA	Hypoaldosteronism; failure to excrete NH_4^+
		Hyperkalemia due to lack of aldosterone
		Normal anion gap
Metabolic alkalosis	Vomiting	Loss of gastric H^+; leaves HCO_3^- behind in blood
		Worsened by volume contraction
		Exhibits hypokalemia
		May have ↑ anion gap due to production of ketoacids (starvation)
	Hyperaldosteronism	Increased H^+ secretion by distal tubule
	Loop or thiazide diuretics	Volume contraction alkalosis
Respiratory acidosis	Opiates; sedatives; anesthetics	Inhibition of medullary respiratory center
	Guillain-Barré syndrome; polio; ALS; multiple sclerosis	Weakening of respiratory muscles
	Airway obstruction	
	Adult respiratory distress syndrome; chronic obstructive pulmonary disease	↓ CO_2 exchange in pulmonary capillaries
Respiratory alkalosis	Pneumonia; pulmonary embolus	Hypoxemia causes ↑ ventilation rate
	High altitude	Hypoxemia causes ↑ ventilation rate
	Psychogenic	
	Salicylate intoxication	Direct stimulation of medullary respiratory center
		Also causes metabolic acidosis

(1) **The serum anion gap is high** if the concentration of "other" anions (e.g., lactate, β-OH butyrate, formate) is increased.

(2) **The serum anion gap is normal** if the concentration of Cl^- is increased (**hyperchloremic metabolic acidosis**).

2. **Metabolic alkalosis**

   a. There is a **decrease in arterial [H^+]** from loss of fixed H^+ or gain of base.

   b. **Arterial [HCO_3^-] increases** because of the loss of H^+.

      −**For example,** in **vomiting,** when H^+ is lost from the stomach, HCO_3^- is left behind in the blood.

   c. **Alkalemia** causes **hypoventilation,** the **respiratory compensation** for metabolic alkalosis.

   d. **Renal compensation** for metabolic alkalosis consists of increased excretion of HCO_3^-, as the filtered load of HCO_3^- exceeds the ability of the renal tubule to reabsorb it.

      −If metabolic alkalosis is accompanied by **ECF volume contraction** (e.g., vomiting), then HCO_3^- reabsorption increases and the metabolic alkalosis worsens.

3. **Respiratory acidosis**

   −is **caused by a primary decrease in respiratory rate** and **retention of CO_2.**

   a. Increased arterial P_{CO_2} causes an **increase in [H^+] and [HCO_3^-]** by mass action.

   b. There is **no respiratory compensation** for respiratory acidosis.

   c. **Renal compensation** consists of increased excretion of H^+ as titratable H^+ and NH_4^+ and increased reabsorption of HCO_3^-, all aided by the increased P_{CO_2}, which supplies more H^+ to the renal cells for secretion. The resulting increase in serum [HCO_3^-] helps to normalize the pH.

4. **Respiratory alkalosis**

   −is **caused by a primary increase in respiratory rate** and **loss of CO_2.**

   a. Decreased arterial P_{CO_2} causes a **decrease in [H^+] and [HCO_3^-]** by mass action.

   b. There is **no respiratory compensation** for respiratory alkalosis.

   c. **Renal compensation** consists of decreased excretion of H^+ as titratable acid and NH_4^+ and decreased reabsorption of HCO_3^-, aided by the decreased P_{CO_2}, which causes a deficit of H^+ in the renal cells for secretion. The resulting decrease in serum [HCO_3^-] helps to normalize the pH.

X. Diuretics (Table 5-10)

XI. Integrative Examples

A. **Hypoaldosteronism**

   1. **Case study**

      −A woman has a history of weakness, weight loss, orthostatic hypotension, increased pulse rate, and increased skin pigmentation. She has decreased serum [Na^+], decreased serum osmolarity, increased serum [K^+], and arterial blood gases consistent with metabolic acidosis.

Table 5-10. Summary of Diuretic Effects on the Nephron

Class of Diuretic	Site of Action	Mechanism	Major Effects
Carbonic anhydrase inhibitors (acetazolamide)	Proximal tubule	Inhibition of carbonic anhydrase	↑ HCO_3^- excretion
Loop diuretics (furosemide, ethacrynic acid, bumetanide)	Thick ascending limb of the loop of Henle	Inhibition of Na^+–K^+–$2Cl^-$ cotransport	↑ NaCl excretion ↑ K^+ excretion (because of ↑ distal tubule flow rate) ↑ Ca^{2+} excretion (treat hypercalcemia) ↓ ability to concentrate urine (because of ↓ corticopapillary gradient) ↓ ability to dilute urine (because of inhibition of diluting segment)
Thiazide diuretics (chlorothiazide, hydrochlorothiazide)	Early distal tubule (cortical diluting segment)	Inhibition of NaCl reabsorption	↑ NaCl excretion ↑ K^+ excretion (because of ↑ distal tubule flow rate) ↓ Ca^{2+} excretion (opposite effect of other diuretics) ↓ ability to dilute urine (because of inhibition of cortical diluting segment) No effect on ability to concentrate urine
K^+-sparing diuretics (spironolactone, triamterene, amiloride)	Late distal tubule and collecting duct	Inhibition of Na^+ reabsorption Inhibition of K^+ secretion (Note: spironolactone is an aldosterone antagonist; the others act directly)	↑ Na^+ excretion (small effect) ↓ K^+ excretion

2. Explanation of hypoaldosteronism

a. The **lack of aldosterone** has direct effects on the kidney—decreased Na^+ reabsorption, decreased K^+ secretion, and decreased H^+ secretion. As a result, there is **ECF volume contraction** (caused by increased Na^+ excretion), **hyperkalemia** (caused by decreased K^+ excretion), and **metabolic acidosis** (caused by decreased H^+ excretion).

b. The ECF volume contraction is responsible for this woman's **orthostatic hypotension**. **Increased pulse rate** is secondary to decreased arterial blood pressure via the baroreceptor mechanism.

c. ECF volume contraction also stimulates **ADH secretion from the posterior pituitary** (via volume receptors). ADH causes increased water reabsorption from the collecting ducts, which causes dilution of the serum [Na^+] and osmolarity. Thus, ADH released by a volume mechanism is "inappropriate" for the serum osmolarity in this case.

d. **Hyperpigmentation** is caused by adrenal insufficiency; a decrease in cortisol levels, by negative feedback, stimulates secretion of adrenocorticotropic hormone (ACTH). ACTH has pigmenting effects similar to those of melanocyte-stimulating hormone.

B. Vomiting

1. Case study

—A man is admitted to the hospital for evaluation of severe epigastric pain. He has had persistent nausea and vomiting for 4 days prior to admission. Upper gastrointestinal (GI) endoscopy demonstrates a pyloric ulcer with partial gastric outlet obstruction. He has orthostatic hypotension, decreased serum [K^+], decreased serum [Cl^-], arterial blood gases consistent with metabolic alkalosis, and decreased ventilation rate.

2. Explanation of responses to vomiting

a. Loss of H^+ from the stomach by vomiting causes increased blood [HCO_3^-] and **metabolic alkalosis**. Because Cl^- is lost from the stomach along with H^+, **hypochloremia** and **ECF volume contraction** result.

b. The decreased ventilation rate is the **respiratory compensation for metabolic alkalosis**.

c. ECF volume contraction causes decreased blood volume and **decreased renal perfusion pressure**. As a result, there is increased secretion of renin, increased production of angiotensin II, and **increased secretion of aldosterone**. The ECF volume contraction also worsens the metabolic alkalosis by increasing HCO_3^- reabsorption in the proximal tubule (contraction alkalosis).

d. The increased levels of aldosterone (secondary to volume contraction) cause increased distal K^+ secretion and **hypokalemia**. Increased aldosterone also causes increased distal H^+ secretion, further worsening the metabolic alkalosis.

e. **Treatment** consists of NaCl infusion to correct ECF volume contraction (which is perpetuating the metabolic alkalosis and causing hypokalemia).

Review Test

Directions: Each of the numbered items or incomplete statements in this section is followed by answers or by completions of the statement. Select the **one** lettered answer or completion that is **best** in each case.

1. K^+ secretion by the distal tubule will be increased by all of the following factors EXCEPT

(A) metabolic alkalosis
(B) high K^+ diet
(C) hyperaldosteronism
(D) spironolactone administration
(E) thiazide diuretic administration

2. Subjects A and B are 70-kg males. Subject A drinks 2 liters of distilled water, and subject B drinks 2 liters of isotonic NaCl. As a result of these ingestions, subject B will have a

(A) greater change in ICF volume
(B) higher positive free-water clearance (C_{H_2O})
(C) greater change in plasma osmolarity
(D) higher urine osmolarity
(E) higher urine flow rate

Questions 3 and 4

A woman with a history of severe diarrhea has the following arterial blood values:

$$pH = 7.25$$
$$P_{CO_2} = 24 \text{ mm Hg}$$
$$[HCO_3^-] = 10 \text{ mEq/L}$$

Venous blood samples show decreased blood $[K^+]$ and a normal anion gap.

3. The correct diagnosis for this patient is

(A) metabolic acidosis
(B) metabolic alkalosis
(C) respiratory acidosis
(D) respiratory alkalosis
(E) normal acid–base status

4. Which of the following statements about this patient is correct?

(A) She is hypoventilating
(B) The decreased arterial $[HCO_3^-]$ is a result of loss of HCO_3^- in diarrheal fluid
(C) The decreased blood $[K^+]$ is a result of exchange of intracellular H^+ for extracellular K^+
(D) The decreased blood $[K^+]$ is a result of decreased circulating levels of aldosterone
(E) The decreased blood $[K^+]$ is a result of decreased circulating levels of ADH

5. Use the values below to answer the following question.

Glomerular capillary hydrostatic pressure	47 mm Hg
Bowman's space hydrostatic pressure	10 mm Hg
Bowman's space oncotic pressure	0 mm Hg

At what value of glomerular capillary oncotic pressure would glomerular filtration stop?

(A) 57 mm Hg
(B) 47 mm Hg
(C) 37 mm Hg
(D) 10 mm Hg
(E) 0 mm Hg

6. Which of the following is a feature of the reabsorption of filtered HCO_3^-?

(A) It results in reabsorption of less than 50% of the filtered load when the plasma concentration of HCO_3^- is 24 mEq/L
(B) It acidifies tubular fluid to pH 4.4
(C) It is directly linked to excretion of H^+ as NH_4^+
(D) It is inhibited by decreases in arterial P_{CO_2}
(E) It can proceed normally in the presence of a renal carbonic anhydrase inhibitor

7. The following information was obtained in a human subject:

Plasma	**Urine**
[Inulin] = 1 mg/ml	[Inulin] = 150 mg/ml
[X] = 2 mg/ml	[X] = 100 mg/ml
	Urine flow rate = 1 ml/min

Assuming that X is freely filtered, which of the following statements is most correct?

(A) There is net secretion of X
(B) There is net reabsorption of X
(C) There is both reabsorption and secretion of X
(D) The clearance of X could be used to measure GFR
(E) The clearance of X is greater than the clearance of inulin

8. To maintain normal H^+ balance, total daily excretion of H^+ should equal the daily

(A) fixed acid production plus fixed acid ingestion
(B) HCO_3^- excretion
(C) HCO_3^- filtered load
(D) titratable acid excretion
(E) filtered load of H^+

9. One gram of mannitol was injected into a female subject. After equilibration, a plasma sample had a mannitol concentration of 0.08 g/L. During the equilibration period, 20% of the injected mannitol was excreted in the urine. The subject's

(A) ECF volume is 1 L
(B) ICF volume is 1 L
(C) ECF volume is 10 L
(D) ICF volume is 10 L
(E) interstitial volume is 12.5 L

10. At plasma concentrations of glucose higher than occur at T_m, the

(A) clearance of glucose is zero
(B) excretion rate of glucose equals the filtration rate of glucose
(C) reabsorption rate of glucose equals the filtration rate of glucose
(D) excretion rate of glucose increases with increasing plasma glucose concentrations
(E) renal vein glucose concentration equals the renal artery glucose concentration

11. All of the following will have a positive C_{H_2O} EXCEPT a

(A) person who drinks 2 L of distilled water in 30 minutes
(B) person who begins excreting large volumes of urine with an osmolarity of 100 mOsm/L following a severe head injury
(C) person receiving lithium treatment for depression, who develops polyuria that is unresponsive to administration of ADH
(D) person with an oat cell carcinoma of the lung, who excretes urine with an osmolarity of 1000 mOsm/L

12. A buffer pair (HA/A^-) has a pK of 5.4. At a blood pH of 7.4, the concentration of HA is

(A) 1/100 that of A^-
(B) 1/10 that of A^-
(C) equal to that of A^-
(D) 10 times that of A^-
(E) 100 times that of A^-

13. Which of the following would produce an increase in reabsorption of isosmotic fluid in the proximal tubule?

(A) Increased filtration fraction
(B) ECF volume expansion
(C) Decreased peritubular capillary protein concentration
(D) Increased peritubular capillary hydrostatic pressure
(E) Oxygen deprivation

14. Which of the following substances or combinations of substances could be used to measure interstitial fluid volume?

(A) Mannitol
(B) D_2O alone
(C) Evans blue
(D) Inulin and D_2O
(E) Inulin and radioactive albumin

15. At plasma PAH concentrations below T_m, PAH

(A) reabsorption is not saturated
(B) clearance equals inulin clearance
(C) secretion rate equals PAH excretion rate
(D) concentration in the renal vein is close to zero
(E) concentration in the renal vein equals PAH concentration in the renal artery

16. Compared with a person who ingests 2 L of distilled water, a person with water deprivation will have a

(A) higher C_{H_2O}
(B) lower plasma osmolarity
(C) lower circulating level of ADH
(D) higher TF/P osmolarity in the proximal tubule
(E) higher rate of H_2O reabsorption in collecting ducts

17. Which of the following would cause an increase in both GFR and RPF?

(A) Hyperproteinemia
(B) A ureteral stone
(C) Dilation of the afferent arteriole
(D) Dilation of the efferent arteriole
(E) Constriction of the efferent arteriole

18. A patient has the following arterial blood values:

pH	7.52
P_{CO_2}	20 mm Hg
$[HCO_3^-]$	16 mEq/L

Which of the following statements about this patient is most likely to be correct?

(A) He is hypoventilating
(B) He has decreased ionized $[Ca^{2+}]$ in blood
(C) He has almost complete respiratory compensation
(D) He has an acid–base disorder caused by overproduction of fixed acid
(E) Appropriate renal compensation would cause his arterial $[HCO_3^-]$ to increase

19. Which of the following would best distinguish an otherwise healthy person with severe water deprivation from a person with SIADH?

(A) C_{H_2O}
(B) Urine osmolarity
(C) Plasma osmolarity
(D) Circulating levels of ADH
(E) Corticopapillary osmotic gradient

20. Which of the following causes a decrease in renal Ca^{2+} clearance?

(A) Hypoparathyroidism
(B) Treatment with chlorothiazide
(C) Treatment with furosemide
(D) ECF volume expansion
(E) Hypermagnesemia

21. A patient arrives in the emergency room with low arterial pressure, reduced tissue turgor, and the following arterial blood values:

pH	7.69
$[HCO_3^-]$	57 mEq/L
P_{CO_2}	48 mm Hg

Which of the following responses would also be expected to occur in this patient?

(A) Hyperventilation
(B) Decreased K^+ secretion by distal tubules
(C) Increased ratio of $H_2PO_4^-$ to HPO_4^{-2} in urine
(D) Exchange of intracellular H^+ for extracellular K^+

22. All of the following ions have a higher concentration in ECF than in ICF EXCEPT

(A) Na^+
(B) K^+
(C) Cl^-
(D) HCO_3^-
(E) Ca^{2+}

23. A patient has a plasma osmolarity of 300 mOsm/L and a urine osmolarity of 1200 mOsm/L. The correct diagnosis is

(A) SIADH
(B) water deprivation
(C) central diabetes insipidus
(D) nephrogenic diabetes insipidus
(E) drinking large volumes of distilled water

24. A patient is infused with PAH to measure RBF. She has a urine flow rate of 1 ml/min, a plasma [PAH] of 1 mg/ml, a urine [PAH] of 600 mg/ml, and a hematocrit of 45%. What is her "effective" RBF?

(A) 600 ml/min
(B) 660 ml/min
(C) 1091 ml/min
(D) 1333 ml/min

25. Which of the following substances has the highest renal clearance?

(A) PAH
(B) Inulin
(C) Glucose
(D) Sodium
(E) Chloride

26. A woman runs a marathon in 90°-weather and replaces all volume lost in sweat by drinking distilled water. After the marathon she will have

(A) decreased TBW
(B) decreased hematocrit
(C) decreased ICF volume
(D) decreased plasma osmolarity
(E) increased intracellular osmolarity

27. Which of the following causes hyperkalemia?

(A) Exercise
(B) Alkalosis
(C) Insulin injection
(D) Decreased serum osmolarity
(E) Treatment with β-agonists

28. All of the following are causes of metabolic acidosis EXCEPT

(A) diarrhea
(B) chronic renal failure
(C) ethylene glycol ingestion
(D) treatment with acetazolamide
(E) hyperaldosteronism

29. All of the following are actions of PTH on the renal tubule EXCEPT

(A) stimulation of adenylate cyclase
(B) stimulation of distal tubule K^+ secretion
(C) stimulation of distal tubule Ca^{2+} reabsorption
(D) inhibition of proximal tubular phosphate reabsorption
(E) stimulation of production of 1,25-dihydroxycholecalciferol

Directions: Each group of items in this section consists of lettered options followed by a set of numbered items. For each item, select the **one** lettered option that is most closely associated with it. Each lettered option may be selected once, more than once, or not at all.

Questions 30–34

Match each numbered patient description below with the correct set of arterial blood values.

	pH	HCO_3^- (mEq/L)	P_{CO_2} (mm Hg)
(A)	7.65	48	45
(B)	7.50	15	20
(C)	7.40	24	40
(D)	7.32	30	60
(E)	7.31	16	33

30. A heavy smoker with a history of emphysema and chronic bronchitis who is becoming increasingly somnolent

31. A patient with partially compensated respiratory alkalosis after 1 month on a mechanical ventilator

32. A patient with chronic renal failure (eating a normal protein diet) and decreased urinary excretion of NH_4^+

33. A patient with untreated diabetes mellitus and increased urinary excretion of NH_4^+

34. A patient with 5-day history of vomiting

Questions 35–38

For the phenomena described in the following questions, choose the appropriate labeled nephron site on the diagram.

35. Nephron site where the amount of K^+ in tubular fluid can exceed the amount of filtered K^+ in a person on a high K^+ diet

36. Site where TF/P osmolarity is lowest in a person who has been deprived of water

37. Site where the tubular fluid inulin concentration is highest during antidiuresis

38. Site of highest tubular fluid glucose concentration

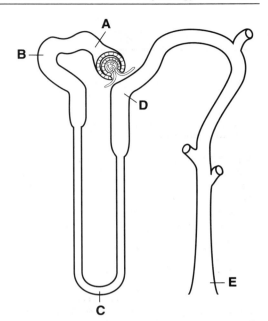

Questions 39–41

The graph below indicates the percentage of the filtered load remaining in tubular fluid at various sites along the nephron. Match each numbered substance with the curve that describes its renal handling.

39. Inulin

40. Alanine

41. PAH

Answers and Explanations

1–D. Distal K^+ secretion is increased by factors that increase the driving force for passive diffusion of K^+ across the luminal membrane. Alkalosis, a diet high in K^+, and hyperaldosteronism all increase $[K^+]$ in the distal cells and thereby increase K^+ secretion. Thiazides increase flow through the distal tubule and dilute the luminal $[K^+]$, so that the driving force for K^+ secretion is increased. Because spironolactone is an aldosterone antagonist, it reduces K^+ secretion.

2–D. After drinking distilled water, subject A will have an increase in ICF and ECF volumes, a decrease in plasma osmolarity, a suppression of ADH secretion, a positive free-water clearance, and will make *dilute* urine with a high flow rate. Subject B, after drinking the same volume of isotonic NaCl, will have an increase in ECF volume only and no change in his plasma osmolarity. Because his ADH will not be suppressed, he will have a higher urine osmolarity, a lower urine flow rate, and a lower C_{H_2O} than subject A.

3–A. An acid pH, together with a decreased HCO_3^- and decreased P_{CO_2}, is consistent with metabolic acidosis with respiratory compensation (hyperventilation). Diarrhea causes GI loss of HCO_3^-, creating a metabolic acidosis.

4–B. The decreased arterial $[HCO_3^-]$ is caused by GI loss of HCO_3^- with diarrhea. She is hyperventilating as respiratory compensation for metabolic acidosis. Her hypokalemia cannot be the result of exchange of intracellular H^+ for extracellular K^+, because she has an increase in extracellular H^+, which would drive the exchange in the other direction. Her circulating levels of aldosterone would be increased as a result of ECF volume contraction.

5–C. Glomerular filtration will stop when the net ultrafiltration pressure across the glomerular capillary is zero; that is, when the force favoring filtration (47 mm Hg) exactly equals the forces opposing filtration (10 mm Hg + 37 mm Hg).

6–D. Decreases in arterial P_{CO_2} cause a decrease in reabsorption of filtered HCO_3^- by diminishing the supply of H^+ in the cell for secretion into the lumen. Reabsorption of filtered HCO_3^- is almost 100% of the filtered load and requires carbonic anhydrase in the brush border to convert filtered HCO_3^- to CO_2 and carbonic anhydrase in the cell to proceed normally. This process causes little acidification of the urine and is not linked to net excretion of H^+ as titratable acid or NH_4^+.

7–B. To answer this question, calculate GFR and C_X. GFR = 150 mg/ml × 1 ml/min / 1 mg/ml = 150 ml/min. C_X = 100 mg/ml × 1 ml/min / 2 mg/ml = 50 ml/min. Since the clearance of X is less than the clearance of inulin (or GFR), then there must have been *net reabsorption* of X. Clearance data alone cannot be used to determine whether there has also been secretion of X. Because GFR cannot be measured with a substance that is reabsorbed, X would not be suitable.

8–A. Total daily production of fixed H^+ from catabolism of proteins and phospholipids (plus any additional fixed H^+ that is ingested) must be matched by the sum of excretion of H^+ as titratable acid plus NH_4^+ in order to maintain acid–base balance.

9–C. Mannitol is a marker substance for the ECF volume. ECF volume = amount of mannitol/ concentration of mannitol = 1 g − 0.2 g / 0.08 g/L = 10 L.

10–D. Above the T_m for glucose, the carriers are saturated so reabsorption rate no longer matches filtration rate—the difference is excreted in the urine. As plasma glucose concentration increases, the excretion of glucose increases. Above T_m, the renal vein glucose concentration will be less than the renal artery concentration, because some glucose is being excreted in urine and is not returned to the blood. The clearance of glucose is zero below T_m (or below threshold) when all the filtered glucose is reabsorbed but is greater than zero above the T_m.

11–D. A person making hyperosmotic urine of 1000 mOsm/L will have a negative C_{H_2O} (C_{H_2O} = $V - C_{osm}$). All of the others will have a positive C_{H_2O} because they are producing hyposmotic urine either from suppression of ADH by water drinking (A), central diabetes insipidus (B), or nephrogenic diabetes insipidus (C).

12–A. Use the Henderson-Hasselbalch equation to calculate the ratio of HA/A⁻:

$$
\begin{aligned}
pH &= pK + \log A^-/HA \\
7.4 &= 5.4 + \log A^-/HA \\
2.0 &= \log A^-/HA \\
100 &= A^-/HA; \text{ or } HA/A^- \text{ is } 1/100
\end{aligned}
$$

13–A. Increasing filtration fraction means that a larger portion of the RPF is filtered across the glomerular capillaries, which causes an increase in the protein concentration and oncotic pressure of the blood leaving the glomerular capillaries. The blood leaving the glomerular capillaries becomes the peritubular capillary blood supply. This increased oncotic pressure in peritubular capillary blood is a driving force *favoring reabsorption* in the proximal tubule. ECF volume expansion, decreased peritubular capillary protein concentration, and increased peritubular capillary hydrostatic pressure all inhibit proximal reabsorption. Oxygen deprivation would also inhibit reabsorption by stopping the Na^+–K^+ pump in the basolateral membranes.

14–E. Interstitial fluid volume is measured indirectly by taking the difference between ECF volume and plasma volume. Inulin, a large fructose polymer that is restricted to the extracellular space, is a marker for ECF volume. Radioactive albumin is a marker for plasma volume.

15–D. At plasma concentrations below the T_m for PAH secretion, PAH concentration in the renal vein is almost zero because the sum of filtration plus secretion of PAH removes virtually all of it from the renal plasma. Thus, the PAH concentration in the renal vein is less than that in the renal artery because most of the PAH is excreted in urine. PAH clearance is greater than inulin clearance because PAH is filtered and secreted; inulin is only filtered.

16–E. The person with water deprivation will have a higher plasma osmolarity and higher circulating levels of ADH. These effects will increase the rate of H_2O reabsorption in the collecting ducts and create a *negative* C_{H_2O}. TF/P osmolarity in the proximal tubule is not affected by ADH.

17–C. Dilation of the afferent arteriole will increase RPF (because renal vascular resistance is decreased) and increase GFR (because glomerular capillary hydrostatic pressure is increased). Dilation of the efferent arteriole will increase RPF but decrease GFR. Constriction of the efferent arteriole will decrease RPF (increased renal vascular resistance) and increase GFR. Both hyperproteinemia (↑ π in glomerular capillaries) and ureteral stone (↑ hydrostatic pressure in Bowman's space) will oppose filtration.

18–B. First, diagnose the acid–base disorder. Alkaline pH, low P_{CO_2}, and low HCO_3^- are consistent with *respiratory alkalosis*. In respiratory alkalosis, there is a decreased $[H^+]$ so less H^+ is bound to negatively charged sites on plasma proteins. As a result, more Ca^{2+} is bound to proteins and, therefore, the *ionized [Ca^{2+}] decreases*. There is no respiratory compensation for primary respiratory disorders. The patient is hyperventilating, which is the cause of the respiratory alkalosis. Appropriate renal compensation would be decreased reabsorption of HCO_3^-, which would cause his arterial [HCO_3^-] to decrease and his blood pH to decrease (become more normal).

19–C. Both individuals will have hyperosmotic urine, a negative C_{H_2O}, a normal corticopapillary gradient, and high circulating levels of ADH. The person with water deprivation will have a high plasma osmolarity, and the person with SIADH will have a low plasma osmolarity (because of dilution by the inappropriate water reabsorption).

20–B. Thiazide diuretics have a unique effect on the distal tubule; they increase Ca^{2+} reabsorption, thereby decreasing Ca^{2+} excretion and Ca^{2+} clearance. Because PTH increases Ca^{2+} reabsorption, the lack of PTH will cause an increase in Ca^{2+} clearance. Furosemide inhibits Na^+ reabsorption in the thick ascending limb, and ECF volume expansion inhibits Na^+ reabsorption in the proximal tubule; at these sites, Ca^{2+} reabsorption is linked to Na^+ reabsorption, and Ca^{2+} clearance would be increased. Because Mg^{2+} competes with Ca^{2+} for reabsorption in the thick ascending limb, hypermagnesemia will cause increased Ca^{2+} clearance.

21–D. First, diagnose the acid–base disorder. Alkaline pH, with increased HCO_3^- and increased Pco_2, is consistent with metabolic alkalosis with respiratory compensation. The low blood pressure and decreased turgor suggest ECF volume contraction. The reduced $[H^+]$ in blood will cause intracellular H^+ to leave cells in exchange for extracellular K^+. The appropriate respiratory compensation is *hypoventilation,* which is responsible for the elevated Pco_2. H^+ excretion in urine will be decreased, so less titratable acid will be excreted. K^+ secretion by the distal tubules will be increased because aldosterone levels will be increased secondary to ECF volume contraction.

22–B. K^+ is the major intracellular cation.

23–B. This patient's plasma and urine osmolarity, taken together, are consistent with water deprivation. The plasma osmolarity is on the high side of normal, stimulating the posterior pituitary to secrete ADH. Secretion of ADH, in turn, acts on the collecting ducts to increase water reabsorption and make hyperosmotic urine. SIADH would also produce hyperosmotic urine, but the plasma osmolarity would be lower than normal because of the excessive water retention. Central and nephrogenic diabetes insipidus and excessive water intake would all result in hyposmotic urine.

24–C. Effective RPF is calculated from the clearance of PAH ($C_{PAH} = U_{PAH} \times V / P_{PAH} = 600$ ml/min). RBF = RPF / 1 − hematocrit = 1091 ml/min.

25–A. The clearance of PAH is highest of all the substances because it is both filtered and secreted. Inulin is only filtered. The other substances are filtered and subsequently reabsorbed and, therefore, will have clearances less than the inulin clearance.

26–D. The woman has a *net loss of NaCl without a net loss of H_2O.* Therefore, her extracellular and plasma osmolarity will be decreased, which will cause water to flow from ECF to ICF. The intracellular osmolarity will also be decreased after the shift of water. TBW will be unchanged because the woman replaced all volume lost in sweat by drinking water. Hematocrit will be increased because of the shift of water from ECF to ICF, and the shift of water into the red blood cells themselves causing their volume to increase.

27–A. Exercise causes a shift of K^+ from cells into blood, resulting in hyperkalemia. Hyposmolarity, insulin, β-agonists, and alkalosis cause a shift of K^+ from blood into cells and result in hypokalemia.

28–E. Metabolic acidosis is caused by overproduction, ingestion (ethylene glycol), or failure to excrete (chronic renal failure) fixed acids or from loss of HCO_3^- (diarrhea, type 2 renal tubular acidosis [RTA]). Note that acetazolamide, a carbonic anhydrase inhibitor, causes type 2 RTA. Hyperaldosteronism causes metabolic alkalosis because the excretion of H^+ is increased.

29–B. Parathyroid hormone does not alter the renal handling of K^+.

30–D. The history strongly suggests chronic obstructive pulmonary disease (COPD) as a cause of respiratory acidosis. Because of the COPD, ventilation rate is decreased and CO_2 is retained. The $[H^+]$ and $[HCO_3^-]$ are increased by mass action. The HCO_3^- is further increased by the renal compensation for respiratory acidosis (increased HCO_3^- reabsorption by the kidney facilitated by the high Pco_2).

31–B. The blood values in respiratory alkalosis show a decreased P_{CO_2} (the cause) and decreased $[H^+]$ and $[HCO_3^-]$ by mass action. The $[HCO_3^-]$ is further decreased by renal compensation for chronic respiratory alkalosis (decreased HCO_3^- reabsorption).

32–E. In chronic renal failure with ingestion of normal amounts of protein, fixed acids will be produced from catabolism of protein. Because the failing kidney does not produce enough NH_4^+ to excrete all of the fixed acid, metabolic acidosis (with respiratory compensation) results.

33–E. Untreated diabetes mellitus results in production of ketoacids, which are fixed acids causing metabolic acidosis. The urinary excretion of NH_4^+ is increased in this patient because there has been an adaptive increase in renal NH_3 synthesis in response to the metabolic acidosis.

34–A. The history of vomiting (in the absence of any other information) indicates loss of gastric H^+ and, as a result, metabolic alkalosis (with respiratory compensation).

35–E. K^+ is secreted by the late distal tubule and collecting ducts. Because this secretion is affected by dietary K^+, a person on a high K^+ diet can secrete more K^+ into the urine than was originally filtered. At all of the other nephron sites, the amount of K^+ in the tubular fluid is either equal to the amount filtered (site *A*) or less than the amount filtered (since K^+ is reabsorbed in the proximal tubule and the loop of Henle).

36–D. A person deprived of water will have high circulating levels of ADH. TF/P osmolarity is 1.0 throughout the proximal tubule regardless of ADH status. In antidiuresis, TF/P osmolarity > 1.0 at site C because of the large corticopapillary osmotic gradient. At site E, TF/P osmolarity > 1.0 because of water reabsorption out of the collecting ducts. At site D, the tubular fluid is diluted by NaCl reabsorption without water reabsorption, and TF/P osmolarity < 1.0.

37–E. Because inulin, once filtered, is neither reabsorbed nor secreted, its concentration in tubular fluid reflects the amount of water remaining in the tubule. In antidiuresis, water is reabsorbed throughout the nephron (except in the thick ascending limb and cortical diluting segment). Thus, the tubular fluid inulin concentration progressively rises along the nephron as water is reabsorbed, and will be highest in the final urine.

38–A. Glucose is extensively reabsorbed in the early proximal tubule by the Na^+–glucose cotransporter. The tubular fluid glucose concentration is highest in Bowman's space before any reabsorption has occurred.

39–C. Once inulin is filtered, it is neither reabsorbed nor secreted. Thus, 100% of the filtered inulin remains in the tubular fluid at each nephron site and in the final urine.

40–A. Alanine, like glucose, is avidly reabsorbed in the early proximal tubule by a Na^+–amino acid cotransporter. Thus, the percentage of the filtered load of alanine remaining in the tubular fluid declines rapidly along the proximal tubule as alanine is reabsorbed back into the blood.

41–D. PAH is an organic acid that is filtered and subsequently secreted by the proximal tubule. The secretion process adds PAH to the tubular fluid; therefore, the amount present by the end of the proximal tubule is greater than was present in Bowman's space.

6

Gastrointestinal Physiology

I. Structure and Innervation of the Gastrointestinal Tract

A. Structure (Figure 6-1)

1. **Epithelial cells**

   —are specialized in different parts of the gastrointestinal (GI) tract for **secretion** or **absorption**.

2. **Muscularis mucosa**

   —Contraction causes a change in the surface area for secretion or absorption.

3. **Circular muscle**

   —Contraction causes a **decrease in diameter** of the lumen of the GI tract.

4. **Longitudinal muscle**

   —Contraction causes **shortening** of a segment of the GI tract.

5. **Submucosal plexus (Meissner's plexus) and myenteric plexus**

   —comprise the **enteric (intrinsic) nervous system** of the GI tract.

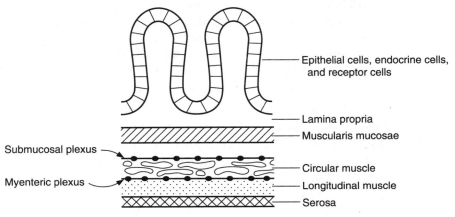

Figure 6-1. Structure of the gastrointestinal tract.

—integrate and coordinate the motility and secretory and endocrine functions of the GI tract.

B. Innervation of the GI tract

—The autonomic nervous system of the GI tract is comprised of both extrinsic and intrinsic nervous systems.

1. Extrinsic innervation (parasympathetic and sympathetic nervous systems)

—**Efferent fibers** carry information from the brainstem and spinal cord to the GI tract.

—**Afferent fibers** carry sensory information (chemoreceptors, mechanoreceptors) from the GI tract back to the brainstem and spinal cord.

a. Parasympathetic nervous system

—is usually **excitatory** on the functions of the GI tract.

—is carried via the **vagus** and **pelvic nerves**.

—Preganglionic parasympathetic fibers **synapse in the myenteric and submucosal plexuses**.

—Cell bodies in ganglia of the plexuses then send information to the smooth muscle, secretory cells, and endocrine cells of the GI tract.

(1) The **vagus nerve** carries information to the esophagus, stomach, pancreas, and upper large intestine.

(2) The **pelvic nerve** carries information to the lower large intestine, rectum, and anus.

b. Sympathetic nervous system

—is usually **inhibitory** on the functions of the GI tract.

—Fibers originate in the spinal cord between T-8 and L-2.

—**Preganglionic cholinergic fibers** synapse in prevertebral ganglia.

—**Postganglionic adrenergic fibers** leave the prevertebral ganglia and **synapse in the myenteric and submucosal plexuses**.

—Direct postganglionic adrenergic innervation of blood vessels and some smooth muscle cells also occurs.

2. Intrinsic innervation (enteric nervous system)

—relays information **to and from the GI tract** via the **parasympathetic and sympathetic nervous systems**.

—relays information **within** the GI tract by **local reflexes**.

—controls most functions of the GI tract, especially motility and secretion.

a. Myenteric plexus (Auerbach's plexus)

—primarily controls the **motility** of the GI smooth muscle.

b. Submucosal plexus (Meissner's plexus)

—primarily controls **secretion and blood flow**.

—receives sensory information from chemoreceptors and mechanoreceptors in the GI tract.

II. Gastrointestinal Hormones

A. Classification

—Four substances have met all requirements to be considered "official" GI hormones; others are considered "candidate" hormones.

1. Hormones

-GI hormones are released into the portal circulation, enter the general circulation, and have physiologic actions at a distant site on target cells with a specific receptor for the hormone.
-The four **GI hormones** are **gastrin, cholecystokinin, secretin,** and **gastric inhibitory peptide**.

2. Paracrines

-are released from endocrine cells in the GI mucosa.
-diffuse short distances to act on target cells within the GI tract.
-Diffusion distance to the target cells must be short, and specific receptors for the paracrine must be present on the target cells.
-The **GI paracrines** are **somatostatin and histamine**.

3. Neurocrines

-are synthesized in neuronal cell bodies, are moved by axonal transport down the axon, and are released by an action potential in the nerve.
-Neurocrines then diffuse over the synaptic cleft to the target cell, where they interact with a specific receptor.
-The GI **neurocrines** are:

a. **Vasoactive intestinal peptide**

b. **Gastrin-releasing peptide (bombesin)**

c. **Enkephalins (met-enkephalin and leu-enkephalin)**

B. GI hormones (Table 6-1)

1. Gastrin

-contains 17 amino acids (**"little gastrin"**).
-"Little gastrin" is the form secreted in response to a meal.
-The **four C-terminal amino acids** contain all of the biologic activity of gastrin.
-**"Big gastrin"** contains 34 amino acids, although it is **not a dimer of "little gastrin."**

a. **Actions of gastrin**
  (1) **Increases gastric H^+ secretion** by the **parietal** (oxyntic) cells of the stomach. Gastrin is far more potent in stimulating gastric H^+ secretion than is histamine.
  (2) **Stimulates growth of gastric oxyntic gland mucosa** and growth of mucosa of the small intestine and colon by stimulating synthesis of RNA and new protein. Patients with **gastrin-secreting tumors** have hypertrophy and hyperplasia of these tissues.

b. **Stimuli for secretion of gastrin**
  -Gastrin is released from the **G cells of the stomach** antrum **in response to a meal**.
  -Gastrin is released in response to the following:
  (1) **Small peptides and amino acids** in the lumen of the stomach. The most potent releasers of gastrin are **phenylalanine** and **tryptophan**.
  (2) **Distention of the stomach**

Table 6-1. Summary of GI Hormones

Hormones	Homology (Family)	Site of Secretion	Stimulus for Secretion	Actions
Gastrin	Gastrin–CCK	G cells of the stomach	Small peptides and amino acids Distention of stomach Vagus (via GRP) Inhibited by H^+ in stomach	↑ gastric H^+ secretion Stimulates growth of gastric mucosa
CCK	Gastrin–CCK	I cells of the duodenum and jejunum	Small peptides and amino acids Fatty acids	Stimulates contraction of gallbladder ↑ pancreatic enzyme and HCO_3^- secretion ↑ growth of exocrine pancreas/gallbladder Inhibits gastric emptying
Secretin	Secretin–glucagon	S cells of the duodenum	H^+ in duodenum Fatty acids in duodenum	↑ pancreatic HCO_3^- secretion ↑ biliary HCO_3^- secretion ↓ gastric H^+ secretion ↓ effect of gastrin on growth of gastric mucosa
GIP	Secretin–glucagon	Duodenum and jejunum	Fatty acids, amino acids and oral glucose	↑ insulin secretion ↓ gastric H^+ secretion

(3) Vagal stimulation mediated by **gastrin-releasing peptide (GRP)**. Atropine does not block vagus-mediated gastrin release because the mediator of the vagal effect is GRP, not acetylcholine (ACh).

(4) **H^+ in the lumen of the stomach inhibits gastrin release.** This negative feedback control ensures that gastrin is not released when the stomach contents are already acidified.

c. **Zollinger-Ellison syndrome**

—occurs when gastrin is secreted by non-beta cell tumors of the pancreas.

2. **Cholecystokinin (CCK)**

—contains 33 amino acids and is chemically **homologous to gastrin**. The five carboxy-terminal amino acids are the same in CCK and gastrin.

—Biologic activity of CCK resides in the C-terminal heptapeptide. The heptapeptide contains the sequence homologous to gastrin and has gastrin activity as well as CCK activity.

a. **Actions of CCK**

(1) Stimulates **contraction of the gallbladder**.

(2) Stimulates **pancreatic enzyme secretion,** and **potentiates the stimulation of pancreatic HCO_3^- secretion by secretin.**

 (3) Stimulates **growth of the exocrine pancreas** and the gallbladder mucosa.
 (4) **Inhibits gastric emptying.** Thus, meals containing fat stimulate the secretion of CCK, which, in turn, slows gastric emptying to allow more time for intestinal digestion and absorption (particularly of fats).

 b. Stimuli for release of CCK
   —CCK is released from the **I cells of the duodenal and jejunal mucosa** by:
   (1) Small peptides and amino acids
   (2) Fatty acids and monoglycerides
     —Triglycerides do not release CCK because they cannot cross the intestinal mucosal cell membrane.

3. Secretin
 —contains 27 amino acids and is chemically **homologous to glucagon;** 14 of the 27 amino acids in secretin are the same as those in glucagon.
 —All of its amino acids are required for activity.

 a. Actions of secretin
   —are coordinated to reduce the amount of H^+ in the lumen of the small intestine.
   (1) Stimulates pancreatic HCO_3^- and H_2O secretion, and increases growth of the exocrine pancreas.
   (2) Stimulates HCO_3^- and H_2O secretion by the liver, and **increases bile production.**
   (3) Inhibits H^+ secretion by the gastric parietal cells, and **inhibits the effect of gastrin on growth of gastric cells.**

 b. Stimuli for release of secretin
   —Secretin is released by the **S cells of the duodenum** in response to:
   (1) H^+ in the lumen of the duodenum
   (2) Fatty acids in the lumen of the duodenum

4. Gastric inhibitory peptide (GIP)
 —contains 42 amino acids and is **homologous to secretin and glucagon.**

 a. Actions of GIP
   (1) Stimulates insulin release. In the presence of an oral glucose load, GIP causes the release of insulin from the pancreas. This action of GIP explains why an **oral glucose load is more effective than intravenous glucose in causing insulin release** and glucose utilization.
   (2) Inhibits H^+ secretion by gastric parietal cells (as is implied in the name).

 b. Stimuli for release of GIP
   —GIP is secreted by the duodenum and jejunum.
   —GIP is the only GI hormone that is released in response to fat, protein, and carbohydrate. Its secretion is stimulated by **fatty acids, amino acids, and orally administered glucose.**

C. GI paracrines

1. Somatostatin

–is secreted by cells throughout the GI tract in response to **H^+ in the lumen**. Its secretion is **inhibited by the vagus** nerve.

–inhibits the release of all GI hormones.

–inhibits gastric H^+ secretion.

2. Histamine

–is secreted by gastric mucosal mast cells.

–**increases gastric H^+ secretion** directly and by potentiating the effects of gastrin and vagal stimulation.

D. GI neurocrines

1. Vasoactive intestinal peptide (VIP)

–contains 28 amino acids and is **homologous to secretin**.

–is released from nerves in the mucosa and smooth muscle of the GI tract.

–produces **relaxation of GI smooth muscle**.

–**stimulates pancreatic HCO_3^- secretion** and **inhibits gastric H^+ secretion**. In these actions, it resembles secretin.

–is also secreted by pancreatic islet cell tumors and is presumed to mediate **pancreatic cholera**.

2. Gastrin-releasing peptide (GRP or bombesin)

–is released from nerves in the gastric mucosa by **vagal stimulation**.

–**stimulates gastrin release**.

3. Enkephalins (met-enkephalin and leu-enkephalin)

–are secreted from nerves in the mucosa and smooth muscle of the GI tract.

–**stimulate contraction of GI smooth muscle,** particularly the lower esophageal, pyloric, and ileocecal sphincters.

–**inhibit intestinal secretion** of fluid and electrolytes. This action forms the basis for the usefulness of **opiates in the treatment of diarrhea**.

III. Gastrointestinal Motility

–Contractile tissue of the GI tract is almost exclusively **unitary smooth muscle. Exceptions** are the pharynx, upper one-third of the esophagus, and the external anal sphincter, all of which are **striated muscle**.

–Depolarization of **circular muscle** leads to contraction of the ring of smooth muscle and a **decrease in diameter** of that segment of the GI tract.

–Depolarization of **longitudinal muscle** leads to contraction in the longitudinal direction and a **decrease in length** of that segment of the GI tract.

–**Phasic contractions** are found in the esophagus, gastric antrum, and small intestine, which contract and relax periodically.

–**Tonic contractions** are found in the lower esophageal sphincter, the orad stomach, and the ileocecal and internal anal sphincters.

A. Slow waves (Figure 6-2)

–are **oscillating membrane potentials** inherent to the smooth muscle cells of some parts of the GI tract.

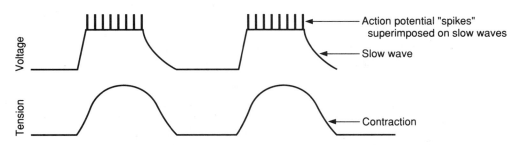

Figure 6-2. Gastrointestinal slow waves superimposed by action potentials. Action potentials produce subsequent contraction.

—are *not* **action potentials,** but they do **determine the pattern of action potentials** and, therefore, the pattern of contraction of the smooth muscle.

1. Mechanism of slow wave production

   —is the cyclic activation and deactivation of the cell membrane Na^+–K^+ pump (Na^+–K^+ adenosine triphosphatase [ATPase]).

   —**Depolarization during each slow wave** brings the membrane potential of smooth muscle cells closer to threshold and, therefore, **increases the probability of action potentials to occur**.

   —The action potentials produced (on top of the background of slow waves) then initiate excitation–contraction coupling and, ultimately, contraction of the smooth muscle cells (see Chapter 1 VII B).

2. Frequency of slow waves

   —varies along the GI tract.

   —is constant and characteristic for each part of the GI tract.

   —**is not influenced** by neural or hormonal input. In contrast, the **frequency of action potentials is** modified by neural and hormonal influences.

   —**The frequency of the slow waves sets the maximum frequency for contraction** of a given part of the GI tract.

   —**is lowest in the stomach** (3 slow waves/min) and **highest in the duodenum** (12 slow waves/min).

B. Chewing, swallowing, and esophageal peristalsis

1. Chewing

   —lubricates food by mixing it with saliva.

   —decreases the size of food particles to facilitate swallowing and to begin the process of digestion.

2. Swallowing

   —The swallowing reflex is **coordinated in the medulla**. Fibers in the vagus and glossopharyngeal nerves carry information to and from the medulla.

   —The mechanism of swallowing is as follows:

   a. The nasopharynx closes and, at the same time, **breathing is inhibited**.

   b. The laryngeal muscles contract to close the glottis and elevate the larynx.

c. Peristalsis begins in the pharynx to propel the food toward the esophagus. Simultaneously, the **upper esophageal sphincter relaxes** to permit entry of the food bolus into the esophagus.

3. Esophageal motility

–The esophagus propels swallowed food into the stomach.

–Sphincters at either end of the esophagus prevent entry of air (upper esophagus) and gastric acid (lower end).

–The upper one-third of the esophagus is **striated muscle**.

–At rest, **intraesophageal pressure is less than atmospheric pressure** (since it is located in the thorax). In fact, intrathoracic pressure can be measured by placing a balloon catheter in the esophagus.

–The mechanism of esophageal motility is as follows:

a. First, the **upper esophageal sphincter relaxes** to permit the swallowed food bolus to enter the esophagus.

b. The upper esophageal sphincter then contracts so that food will not reflux.

c. A **primary peristaltic contraction** creates an area of increased pressure just behind the food bolus.

d. The peristaltic contraction moves down the esophagus, propelling the food bolus along. **Gravity** accelerates the movement.

e. A **secondary peristaltic contraction** clears the esophagus of any food remaining.

f. The **lower esophageal sphincter relaxes** as the food bolus approaches it. This relaxation is vagus-mediated, and the neurotransmitter is **VIP**. At the same time, the orad region of the stomach relaxes (**"receptive relaxation"**), allowing the food bolus to enter the stomach.

4. Clinical correlations of esophageal motility

a. **Gastric reflux (heartburn)** occurs when there is decreased tone of the lower esophageal sphincter (gastric contents reflux into esophagus), or if secondary peristalsis does not completely clear the esophagus of food.

b. **Achalasia** occurs when the lower esophageal sphincter does not relax during swallowing, and food accumulates in the esophagus.

C. Gastric motility

–The stomach has three layers of smooth muscle—the usual longitudinal and circular layers, and a third oblique layer.

–The **orad region** of the stomach, which contains the oxyntic glands, is responsible for receiving the ingested meal.

–The **caudad region** of the stomach is responsible for the contractions that mix food and propel it into the duodenum.

1. "Receptive relaxation"

–The **orad region of the stomach relaxes** to accommodate the ingested meal. This **vagovagal reflex** is initiated by distention of the stomach and abolished by vagotomy.

–**CCK** participates in receptive relaxation by increasing distensibility of the orad region of the stomach.

2. Mixing and digestion

–The **caudad region of the stomach contracts to mix the food with gastric secretions** and begins the process of digestion. The size of food particles is further reduced.

a. Slow waves in the caudad stomach occur at a frequency of 3–5 waves/ min. They depolarize the smooth muscle cells.

b. If threshold is reached during the slow waves, action potentials are fired, followed by contraction. Thus, the frequency of slow waves sets the maximal frequency of contraction.

c. The wave of contraction closes the distal antrum. As the caudad stomach contracts, **food is propelled back into the stomach to be mixed**.

d. Gastric contractions are **increased by vagal stimulation** and **decreased by sympathetic stimulation**.

e. Even during fasting, contractions (called the **migrating myoelectric complex**) occur at 90-minute intervals and clear the stomach of any residual food. **Motilin** causes these contractions.

3. Gastric emptying

–The caudad region of the stomach contracts to propel food into the duodenum.

a. The rate of **gastric emptying is fastest if the stomach contents are isotonic**. Hypertonic and hypotonic contents empty more slowly.

b. Fat inhibits gastric emptying (i.e., increases gastric emptying time) by causing the release of CCK.

c. H^+ in the duodenum inhibits gastric emptying by a direct neural reflex. H^+ receptors in the duodenum relay information to the gastric smooth muscle via interneurons in the GI plexuses.

D. Small intestinal motility

–The **small intestine functions in digestion and absorption** of nutrients. Therefore, the motility of the small intestine serves to mix and expose nutrients to digestive enzymes and to the absorptive mucosa, then to propel the "leftovers" to the large intestine.

–As in the stomach, **slow waves set the basic electrical rhythm** but occur at a frequency of 12 waves/min. Action potentials occur on top of the slow waves and produce contractions.

–**Parasympathetic stimulation** increases intestinal smooth muscle contraction, and **sympathetic stimulation** decreases it.

1. Segmentation contractions

–serve to **mix the contents of the small intestine**.

–A section of small intestine contracts in isolation, sending the contents in both orad and caudad directions. When that section of small intestine relaxes, the contents flow back into the segment.

–This back-and-forth movement produced by the segmentation contractions causes mixing but no net forward movement of the chyme down the small intestine.

2. Peristaltic contractions

–are highly coordinated and serve to **propel the contents down the small intestine** toward the large intestine. This process should occur once adequate digestion and absorption have taken place.

–**Contraction occurs behind the bolus,** and **relaxation occurs in front of the bolus,** causing it to be propelled in the caudad direction.

–The **peristaltic reflex** is coordinated by the **enteric nervous system**.

3. Gastroileal reflex

–is mediated by the extrinsic **autonomic nervous system** and possibly by **gastrin**.

–The presence of food in the stomach triggers increased peristalsis in the ileum and relaxation of the ileocecal sphincter. As a result, intestinal contents are delivered to the large intestine.

E. Large intestinal motility

–Fecal material moves from the cecum to the colon (i.e., ascending, transverse, descending, and sigmoid colons), to the rectum, and then to the anal canal.

–**Haustra,** sac-like segments, appear following contractions of the large intestine.

1. Cecum and proximal colon

–Once the proximal colon is distended with fecal material delivered from the ileum, the ileocecal sphincter contracts, preventing reflux back into the ileum.

a. Segmentation contractions in the proximal colon serve to mix the contents and are responsible for the appearance of **haustra**.

b. Mass movements of large intestinal contents occur 1–3 times/day, whereby the colonic contents are moved distally for long distances (e.g., from the transverse colon to the sigmoid colon).

2. Distal colon

–Because most colonic water absorption takes place in the proximal colon, the fecal material in the distal colon becomes semisolid and moves slowly. Mass movements propel it into the rectum.

3. Rectum, anal canal, and defecation

–The mechanism of defecation is as follows:

a. When the rectum fills with fecal material, it contracts and the internal anal sphincter relaxes (**rectosphincteric reflex**).

b. Once the rectum is filled to about 25% of its capacity, there is an **urge to defecate**. Defecation is prevented because the external anal sphincter is tonically contracted.

c. When it is convenient to defecate, the external anal sphincter is relaxed voluntarily, and the smooth muscle of the rectum contracts to create pressure, forcing the feces out of the body.

–Intra-abdominal pressure may be increased by expiring against a closed glottis (**Valsalva maneuver**).

4. Gastrocolic reflex

−The presence of **food in the stomach** increases motility of the colon and **increases the frequency of mass movements**.

  a. A rapid **parasympathetic** component of the gastrocolic reflex is initiated when the stomach is stretched by food.

  b. A slower, hormonal component is mediated by **CCK and gastrin,** which increase colonic motility.

5. Disorders of large intestinal motility

  a. Emotional factors strongly influence large intestinal motility via the extrinsic autonomic nervous system. **Irritable bowel syndrome** may occur during periods of stress, resulting in constipation (increased segmentation contractions) or diarrhea (decreased segmentation contractions).

  b. Megacolon (Hirschsprung's disease), the **absence of the enteric nervous system of the colon,** results in constriction of the involved segment, marked dilatation and accumulation of intestinal contents proximal to the constriction, and severe constipation.

F. Vomiting

−A wave of reverse peristalsis begins in the small intestine, moving the GI contents in the orad direction.

1. Retching

−occurs when the gastric contents are pushed into the esophagus while the upper esophageal sphincter remains closed.

2. Vomiting

−occurs if the pressure in the esophagus becomes high enough to open the upper esophageal sphincter.

3. Vomiting center in the medulla

−is stimulated by tickling the back of the throat, gastric distention, and vestibular stimulation (motion sickness).

4. Chemoreceptor trigger zone in the fourth ventricle

−is activated by emetics, radiation, and vestibular stimulation.

IV. Gastrointestinal Secretion (Table 6-2)

A. Salivary secretion

1. Functions of saliva

  a. Initial digestion of starch by α-amylase (ptyalin), and **initial digestion of triglycerides** by lingual lipase

  b. Lubrication of ingested food by mucus

  c. Protection of the mouth and esophagus by dilution and buffering of ingested foods.

2. Composition of saliva (Figure 6-3)

  a. Saliva is characterized by:

   (1) High volume (for the size of the salivary glands)

   (2) High K^+ and HCO_3^- concentrations

Table 6-2. Summary of GI Secretions

GI Secretion	Major Characteristics	Stimulated by	Inhibited by
Saliva	High HCO_3^- High K^+ Hypotonic α-Amylase Lingual lipase	Parasympathetic Sympathetic	Sleep Dehydration Atropine
Gastric	HCl	Gastrin Parasympathetic Histamine	↓ stomach pH Chyme in duodenum (via secretin and GIP) Atropine Cimetidine Omeprazole
	Pepsinogen Intrinsic factor	Parasympathetic	
Pancreatic	High HCO_3^- Isotonic	Secretin CCK (potentiates secretin) Parasympathetic (potentiates secretin)	
	Pancreatic lipase, amylase, proteases	CCK Parasympathetic	
Bile	Bile salts Bilirubin Phospholipids, cholesterol	CCK (causes contraction of gallbladder and relaxation of sphincter of Oddi) Parasympathetic (causes contraction of gallbladder)	Ileal resection

(3) Low Na^+ and Cl^- concentrations
(4) Hypotonicity
(5) α-Amylase, lingual lipase, and kallikrein

b. The composition of saliva varies with flow rate (see Figure 6-3).

(1) At lowest flow rates, saliva has the lowest osmolarity and lowest Na^+, Cl^-, and HCO_3^- concentrations. It has the highest K^+ concentration.

(2) At highest flow rates (up to 4 ml/min), the composition of saliva is closest to that of plasma.

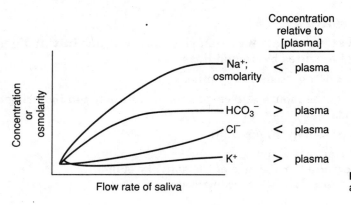

	Concentration relative to [plasma]
Na^+; osmolarity	< plasma
HCO_3^-	> plasma
Cl^-	< plasma
K^+	> plasma

Figure 6-3. Composition of saliva as a function of salivary flow rate.

3. Formation of saliva (Figure 6-4)

–Saliva is formed by three major glands: **parotid, submaxillary, and sublingual glands**.

–The **structure** of each gland is similar to a "bunch of grapes." An **acinus,** the blind end of each duct, is lined with acinar cells and secretes the initial saliva. The **branching duct system** is lined with columnar epithelial cells, which modify the saliva.

–**Myoepithelial cells** in the acinus and initial ducts contract when saliva production is stimulated and eject saliva into the mouth.

a. The acinus

–**produces an initial saliva** whose composition is about the same as **plasma**.

–This fluid is **isotonic** and has the same Na^+, K^+, Cl^-, and HCO_3^- concentrations as plasma.

b. The ducts

–**modify the saliva** by the following processes:

(1) They **reabsorb Na^+ and Cl^-,** which makes the concentrations of these ions lower than plasma concentrations.

(2) They **secrete K^+ and HCO_3^-,** which makes the concentrations of these ions higher than plasma concentrations.

(3) **Aldosterone** acts on the ductal cells to increase the reabsorption of Na^+ and the secretion of K^+ (analogous to its action on the renal distal tubule and colon).

(4) The **saliva becomes hypotonic** because the ducts are relatively impermeable to water. Since more solute than water is reabsorbed, the saliva becomes dilute relative to plasma.

(5) The **effect of increasing flow rate** on saliva composition can be explained by the decreased time available for reabsorption and secretion processes to occur. Thus, at **high rates of flow,** the saliva will more closely resemble the initial secretion from the acinus and the plasma.

4. Regulation of saliva formation

–Production of saliva is controlled by the parasympathetic and sympathetic nervous systems, not by the GI hormones.

–Saliva production is unique in that it is **increased by both the parasympathetic and sympathetic activity**. Parasympathetic activity is more important.

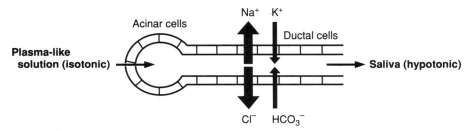

Figure 6-4. Modification of saliva by ductal cells.

a. Parasympathetic stimulation (cranial nerves VII and IX)

—**increases production of saliva** by increasing the transport processes of the acinar and ductal cells and by causing vasodilation.

—The cholinergic receptors are **muscarinic,** and the second messenger is **inositol triphosphate (IP$_3$)** and **increased intracellular [Ca^{2+}]**.

—Anticholinergic drugs (e.g., **atropine**) inhibit the production of saliva and cause **dry mouth**.

b. Sympathetic stimulation

—**increases production of saliva** and growth of salivary glands, although the effects are smaller than those of parasympathetic stimulation.

—The receptors are **β-adrenergic receptors** and the second messenger is **cyclic adenosine monophosphate (AMP)**.

c. Saliva production

—**is increased** (via activation of the parasympathetic nervous system) by food in the mouth, smells, conditioned reflexes, and nausea.

—**is decreased** (via inhibition of the parasympathetic nervous system) by sleep, dehydration, fear, and anticholinergic drugs.

B. Gastric secretion

1. Gastric cell types and their secretions (Table 6-3 and Figure 6-5)

2. Mechanism of gastric H$^+$ secretion (Figure 6-6)

—Parietal cells **secrete HCl into the lumen of the stomach** and, concurrently, **absorb HCO$_3^-$ into the bloodstream**.

a. CO$_2$ and H$_2$O are converted to H$^+$ and HCO$_3^-$ in the parietal cells, catalyzed by **carbonic anhydrase**.

Table 6-3. Gastric Cell Types and Their Functions

Cell Type	Part of Stomach	Secretion Products	Stimulus for Secretion
Oxyntic (parietal) cells	Body (fundus)	HCl	Gastrin Vagal stimulation (ACh) Histamine
		Intrinsic factor (necessary for life)	
Chief (peptic) cells	Body (fundus)	Pepsinogen (converted to pepsin at low pH)	Vagal stimulation (ACh)
G cells	Antrum	Gastrin	Vagal stimulation (via GRP) Small peptides Somatostatin inhibits H$^+$ in the stomach inhibits (via stimulation of somatostatin release)
Mucous cells	Antrum	Mucus Pepsinogen	Vagal stimulation (ACh)

Fundus

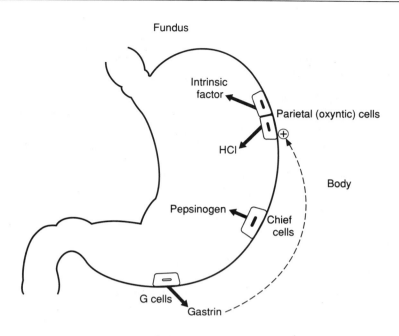

Figure 6-5. Gastric cell types and their functions.

b. The **H^+ is secreted into the lumen of the stomach** by the H^+–K^+ pump (**H^+–K^+ ATPase**). Because Cl^- is secreted along with H^+, the secretion product of the parietal cells is HCl.

   —The drug **omeprazole** is useful in the **treatment of ulcers** because it inhibits the H^+–K^+ ATPase and blocks H^+ secretion.

c. The **HCO_3^- produced in the cells is absorbed into the bloodstream** in exchange for Cl^- (**Cl^-–HCO_3^- exchange**). As HCO_3^- is added to the venous blood, the pH rises (**"alkaline tide"**). (Eventually, this HCO_3^- will be secreted back into the lumen of the GI tract in the pancreatic secretions to neutralize H^+ in the small intestine.)

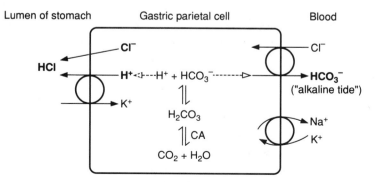

Figure 6-6. Simplified mechanism of secretion of H^+ by gastric parietal cells.

d. If vomiting occurs and there is no H^+ to neutralize in the small intestine, then arterial blood becomes alkaline (**metabolic alkalosis**) because of the HCO_3^- added to it by the parietal cells.

3. Stimulation of gastric H^+ secretion (Figure 6-7)

a. Parasympathetic stimulation

–ACh stimulates H^+ secretion by activating **muscarinic receptors** on the parietal cell membrane.

–The second messenger for ACh is **IP_3 and increased intracellular $[Ca^{2+}]$**.

–**Atropine**, a muscarinic blocking agent, inhibits H^+ secretion by blocking the stimulatory effect of ACh. Atropine is most useful in combination with H_2 receptor-blocking drugs because of the potentiating effect of ACh on histamine action.

b. Histamine

–is released from mast cells in the mucosa and diffuses to the parietal cells.

–stimulates H^+ secretion by activating **H_2 receptors** on the parietal cell membrane.

–The second messenger for histamine is **cyclic AMP**.

–H_2 receptor-blocking drugs such as **cimetidine** inhibit H^+ secretion by blocking the stimulatory effect of histamine.

c. Gastrin

–is, as previously noted, released in response to eating a meal (small peptides, distention of the stomach, vagal stimulation).

–The receptor for gastrin on the parietal cell has not been identified.

–The second messenger for gastrin on the parietal cell has not been identified either, but clearly it is different from those for ACh and histamine since their actions are additive with those of gastrin.

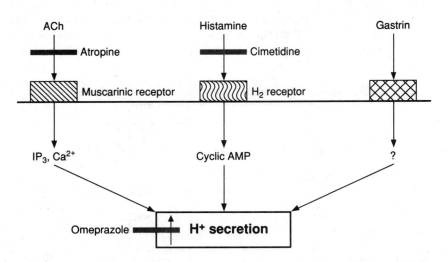

Figure 6-7. Agents and second messengers that alter H^+ secretion by gastric parietal cells.

d. Potentiating effects of ACh, histamine, and gastrin on H^+ secretion

—**Potentiation** occurs when the response to simultaneous administration of two stimulants is greater than the sum of responses to either agent given alone. Thus, low concentrations of stimulants given together can produce maximal effects.

—The potentiation of gastric H^+ secretion is explained, in part, by **each agent having a different mechanism of action** on the parietal cell.

(1) Histamine potentiates the actions of ACh and gastrin in stimulating H^+ secretion. Thus, H_2 receptor blockers (e.g., **cimetidine**) are particularly effective because they block not only the action of histamine but also histamine's potentiating effects on ACh and gastrin.

(2) ACh potentiates the actions of histamine and gastrin in stimulating H^+ secretion. Thus, muscarinic receptor blockers such as **atropine** not only block the action of ACh, but they also block the potentiating effects of ACh on histamine and gastrin.

4. Inhibition of gastric H^+ secretion

—**Negative feedback** mechanisms inhibit the secretion of H^+ by the parietal cells when the pH of the stomach is low.

a. Low pH (< 3.0) in the stomach

—**inhibits gastrin secretion** and thereby inhibits H^+ secretion.

—After a meal is ingested, H^+ secretion is turned on by the mechanisms discussed previously. Once the meal is digested and the stomach emptied, further H^+ secretion lowers the pH of the stomach contents. When the pH of the stomach contents is < 3.0, gastrin release is inhibited and, by negative feedback, further H^+ secretion is inhibited.

b. Chyme in the duodenum

—**inhibits H^+ secretion** either directly or via hormonal mediators.

—The mediators are **GIP** (released by fatty acids in the duodenum) and **secretin** (released by H^+ in the duodenum).

5. Pathophysiology of gastric H^+ secretion

a. Gastric ulcers

—When the usual protective permeability barrier of the stomach is impaired, the presence of H^+ and pepsin injures the gastric mucosa.

—**H^+ secretion is lower than normal,** not higher than normal (as might be assumed).

—**Gastrin levels are high** in persons with gastric ulcer disease because of the lower-than-normal levels of H^+ secretion.

b. Duodenal ulcers

—are more common than gastric ulcers.

—**H^+ secretion is higher than normal** and is responsible, along with pepsin, for damaging the duodenal mucosa.

—**Gastrin levels in response to a meal are very high,** and the number of **gastric parietal cells is increased** (because of the trophic effect of gastrin).

  c. Zollinger-Ellison syndrome
   —A **gastrin-secreting tumor of the pancreas** causes increased H^+ secretion.
   —H^+ secretion continues unabated because the gastrin secreted by the pancreatic tumor cells is not subject to negative feedback inhibition by H^+.

6. Drugs that block H^+ secretion are used in the **treatment of ulcers** (see Figure 6-7).

 a. Atropine
  —blocks H^+ secretion by inhibiting cholinergic muscarinic **receptors**, thereby inhibiting ACh stimulation of H^+ secretion.
  —**Vagotomy** deprives the parietal cells of their parasympathetic innervation and inhibits H^+ secretion.

 b. Cimetidine
  —blocks the H_2 receptor, thereby inhibiting histamine stimulation of H^+ secretion.
  —is particularly effective because it not only blocks the histamine stimulation of H^+ secretion, but it also blocks histamine's potentiation of the ACh effect.

 c. Omeprazole
  —blocks the H^+–K^+ ATPase directly and inhibits H^+ secretion.

C. Pancreatic secretion
 —contains HCO_3^-, whose function is to neutralize the contents of the duodenum.
 —contains **enzymes** essential for digestion of protein, carbohydrate, and fats.

1. Composition of pancreatic secretion (Figure 6-8)

 a. Pancreatic juice is characterized by:
  (1) High volume
  (2) Virtually the same Na^+ and K^+ concentrations as plasma
  (3) Much **higher HCO_3^- concentration than plasma**
  (4) Much lower Cl^- concentration than plasma
  (5) Isotonicity
  (6) Pancreatic lipase, amylase, and proteases

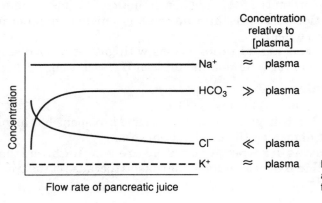

Figure 6-8. Composition of pancreatic juice as a function of pancreatic flow rate.

b. The composition of the aqueous component of pancreatic secretions varies with flow rate.

    (1) Na^+ and K^+ concentrations and osmolarity of pancreatic juice are not affected by flow rate.

    (2) At lowest flow rates, pancreatic juice has the lowest HCO_3^- and the highest Cl^- concentrations.

    (3) At highest flow rates, pancreatic juice has the highest HCO_3^- and the lowest Cl^- concentrations.

2. Formation of pancreatic secretion (Figure 6-9)

    –Like the salivary glands, the exocrine pancreas is also structured like a "bunch of grapes."

    –The acinar cells of the exocrine pancreas make up most of its weight.

    a. The acinus

        –produces a small volume of initial pancreatic juice, which is mostly Na^+ and Cl^-.

    b. Ductal cells

        –modify the initial pancreatic juice by **secreting HCO_3^- and absorbing Cl^- via a Cl^-–HCO_3^- exchange** mechanism in the luminal membrane of the ductal cells.

        –Because the pancreatic ducts **are permeable to water,** H_2O moves into the lumen to make the pancreatic juice isosmotic.

3. Regulation of pancreatic secretion

    a. Secretin

        –is secreted by S cells of the duodenum in response to H^+ in the lumen.

        –**acts on the pancreatic ductal cells to increase HCO_3^- secretion.** Thus, when H^+ is delivered from the stomach to the duodenum, secretin is released, which causes secretion of HCO_3^- into the duodenal lumen to neutralize the H^+.

        –The second messenger for secretin is **cyclic AMP**.

    b. CCK

        –is secreted by I cells of the duodenum in response to small peptides, amino acids, and fatty acids in the lumen.

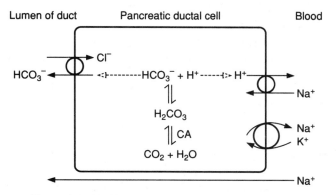

Figure 6-9. Modification of pancreatic juice by ductal cells.

–**acts on the pancreatic acinar cells to increase enzyme secretion** (amylase, lipase, protease).

–potentiates the effect of secretin on the ductal cells to stimulate HCO_3^- secretion.

–The second messenger for CCK is **IP_3 and increased intracellular $[Ca^{2+}]$**. The potentiating effects of CCK on secretin are explained by different mechanisms of action for the two GI hormones.

c. ACh (via vagovagal reflexes)

–is released in response to H^+, small peptides, amino acids, and fatty acids in the duodenal lumen.

–**stimulates enzyme secretion** by the acinar cells, and, like CCK, potentiates the effect of secretin on HCO_3^- secretion.

4. Cystic fibrosis

–is a **disorder of pancreatic secretion**.

–results from a defect in Cl^- channels caused by a mutation in the **cystic fibrosis transmembrane conductance regulator (CFTR) gene**.

–is a **deficiency of pancreatic enzymes,** which causes malabsorption and steatorrhea.

D. Bile secretion and gallbladder

1. Composition and function of bile

–Bile contains **bile salts,** phospholipids, cholesterol, and bile pigments (bilirubin).

a. Bile salts

–are **amphipathic** molecules, having both hydrophilic and hydrophobic portions. In aqueous solution, bile salts orient themselves around droplets of lipid and keep the lipid dispersed in solution (emulsified).

–aid in the intestinal digestion and absorption of lipids by emulsifying and solubilizing them in **micelles**.

b. Micelles

–Above the **critical micellar concentration,** bile salts form micelles.

–Bile salts are on the outside of the micelle, pointing their hydrophilic portions into the aqueous solution of the intestinal lumen.

–Free fatty acids and monoglycerides are on the inside of the micelle, essentially "solubilized" for absorption.

2. Formation of bile

–Bile is **produced continuously by hepatocytes**.

–Bile drains into the hepatic ducts and is stored in the gallbladder for release in response to a meal.

–**Choleretic agents** increase the formation of bile.

–Bile is formed by the following process:

a. Primary **bile acids (cholic acid** and **chenodeoxycholic acid)** are synthesized by hepatocytes. They are conjugated with glycine or taurine to form their respective **bile salts**.

–In the intestine, bacteria convert a portion of each of the primary bile acids to **secondary bile acids (deoxycholic acid** and **lithocholic acid)**.

−Synthesis occurs, as needed, to replace bile acids that are excreted in the feces rather than recirculated back to the liver.

b. Electrolytes and H_2O are added to the bile.

c. The gallbladder fills with bile during the interdigestive period when it is relaxed and the sphincter of Oddi is contracted.

d. The **gallbladder concentrates the bile** by reabsorbing Na^+, Cl^-, and HCO_3^-. H_2O is reabsorbed isosmotically.

3. Contraction of the gallbladder

a. CCK

−**is released** in response to small peptides and fatty acids in the duodenum.

−tells the gallbladder that fats need to be emulsified and absorbed—in other words, bile is needed.

−causes the following:

(1) Contraction of the gallbladder smooth muscle

(2) Relaxation of the sphincter of Oddi

b. ACh

−also causes contraction of the gallbladder.

4. Recirculation of bile acids to the liver

−The **terminal ileum** contains a mechanism for **secondary active transport** of conjugated bile acids with Na^+, which recirculates bile acids to the liver.

−Most of the bile acids are not recirculated to the liver until they reach the terminal ileum. This ensures that bile acids will be present for maximal absorption of fats throughout the upper small intestine.

−**Ileal resection causes steatorrhea**: Bile acids lost in feces are not recirculated to the liver, and the bile acid pool becomes depleted.

V. Digestion and Absorption (Table 6-4)

−Carbohydrates, protein, and lipids are digested and absorbed in the small intestine.

−The surface area for absorption in the small intestine is greatly increased by the presence of the **brush border**.

A. Carbohydrates

1. Digestion of carbohydrates

−**Only monosaccharides are absorbed,** and the carbohydrates must be digested to glucose, galactose, and fructose for absorption to proceed.

a. α-Amylases (salivary and pancreatic) hydrolyze 1,4-glycosidic bonds in starch, yielding maltose, maltotriose, and α-limit dextrins.

b. Maltase, α-dextranase, and sucrase in the intestinal brush border hydrolyze the oligosaccharides to the final monosaccharide products—glucose, galactose, and fructose.

c. Lactase, trehalase, and sucrase degrade their respective disaccharides to monosaccharides.

Table 6-4. Summary of Digestion and Absorption

Nutrient	Digestion	Site of Absorption	Mechanism of Absorption
Carbohydrates	To monosaccharides (glucose, galactose, fructose)	Small intestine	Na^+—dependent cotransport (glucose, galactose) Facilitated diffusion (fructose)
Proteins	To amino acids, dipeptides, tripeptides	Small intestine	Na^+—dependent cotransport
Lipids	To fatty acids, monoglycerides, cholesterol	Small intestine	Micelles form with bile salts in intestinal lumen Diffusion of fatty acids, monoglycerides, and cholesterol into cell Reesterification in the cell to triglycerides and phospholipids Chylomicrons form in the cell (requires apoprotein) and are transferred to lymph
Fat-soluble vitamins		Small intestine	Micelles with bile salts
Water-soluble vitamins		Small intestine	Na^+—dependent cotransport
Vitamin B_{12}		Ileum of small intestine	Intrinsic factor
Bile acids		Ileum of small intestine	Na^+—dependent cotransport; recirculated back to the liver
Ca^{2+}		Small intestine	Vitamin D—dependent Ca^{2+}—binding protein
Fe^{2+}	Fe^{3+} is reduced to Fe^{2+}	Small intestine	Binds to apoferritin in the cell Circulates in blood bound to transferrin

2. Absorption of carbohydrates (Figure 6-10)

a. Glucose and galactose

—are absorbed from the intestinal lumen into the cells by **Na^+—dependent secondary active transport** (cotransport) in the luminal membrane. The sugar is transported uphill and Na^+ is transported downhill.

—are subsequently transported from cell to blood by facilitated diffusion.

—The Na^+–K^+ pump in the basolateral membrane keeps the intracellular $[Na^+]$ low, thus maintaining the Na^+ gradient across the luminal membrane.

—Poisoning the Na^+–K^+ pump will inhibit glucose and galactose absorption because the Na^+ gradient dissipates.

b. Fructose

—is absorbed exclusively by facilitated diffusion; therefore, it cannot be absorbed against a concentration gradient.

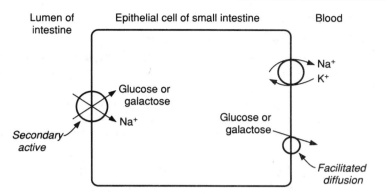

Figure 6-10. Mechanism of absorption of monosaccharides by intestinal epithelial cells. Glucose and galactose are absorbed by Na^+–dependent cotransport (secondary active), and fructose (not shown) is absorbed by facilitated diffusion.

3. Clinical disorders of carbohydrate absorption

–**Lactose intolerance** results from the absence of brush border lactase and thus the inability to hydrolyze lactose to glucose for absorption. Lactose remains in the GI tract as a nonabsorbed solute. H_2O remains in the GI tract isosmotically and causes **osmotic diarrhea**.

B. Protein

–can be **absorbed as amino acids, dipeptides, and tripeptides** (in contrast to carbohydrates, which can only be absorbed as monosaccharides).

1. Digestion of proteins

a. Endopeptidases

–degrade proteins by hydrolyzing interior peptide bonds.

b. Exopeptidases

–hydrolyze one amino acid at a time from the C terminus of proteins and peptides.

c. Pepsin

–is secreted as pepsinogen by the chief cells of the stomach.
–is not essential for protein digestion.
–Pepsinogen is activated to pepsin by gastric H^+.
–The **pH optimum for pepsin is between 1 and 3**. Pepsin is denatured when the pH is > 5 and is inactivated in the duodenum where HCO_3^- is secreted.

d. Pancreatic proteases

–are secreted in an inactive form and are activated by brush border enzymes.
–**For example,** trypsinogen is secreted by the pancreas and is activated to trypsin by enterokinase in the small intestine.
–**Pancreatic proteases degrade each other** and are absorbed along with dietary proteins.

2. Absorption of protein (Figure 6-11)

a. Absorption of free amino acids

—There is **Na⁺–dependent secondary active transport** (cotransport) of amino acids in the luminal membrane, analogous to the transporter for glucose and galactose.

—The amino acids are subsequently transported from cell to blood by facilitated diffusion and simple diffusion.

—There are **four separate carriers** for neutral, acidic, basic, and imino amino acids, respectively.

b. Absorption as dipeptides and tripeptides

—is **faster than absorption of free amino acids**.

—There is **Na⁺–dependent secondary active transport** for dipeptides and tripeptides in the luminal membrane.

—Once the dipeptides and tripeptides have been transported into the intestinal cells, cytoplasmic peptidases hydrolyze them to amino acids. They are then transported from cell to blood by facilitated and simple diffusion.

C. Lipids

—are absorbed into intestinal cells as fatty acids, monoglycerides, and cholesterol. Once inside the intestinal cells, these are resynthesized to triglycerides and phospholipids and then transported in the lymph in chylomicrons.

1. Digestion of lipids in the lumen of the GI tract

a. Stomach

(1) Mixing in the stomach **breaks lipids into droplets** to increase the surface area for digestion by pancreatic enzymes.

(2) **Lingual lipases** digest some of the lipid; however, most of the ingested lipids will be digested in the intestine by pancreatic lipases.

(3) **Gastric emptying is slowed by CCK;** the release of lipids from the stomach to the duodenum is slow enough for adequate digestion and absorption.

b. Small intestine

(1) The components of **bile** help to emulsify lipids in the small intestine.

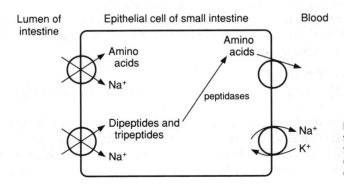

Figure 6-11. Mechanism of absorption of amino acids, dipeptides, and tripeptides by intestinal epithelial cells. Each is absorbed by Na⁺–dependent cotransport.

(2) **Pancreatic enzymes** hydrolyze lipids to fatty acids, monoglycerides, and cholesterol. The enzymes are:
  (a) Pancreatic lipase
  (b) Cholesterol ester hydrolase
  (c) Pancreatic phospholipase A_2
(3) **Micelles** form and "solubilize" the hydrophobic products of lipid digestion.

2. Absorption of lipids

  a. Micelles bring the products of lipid digestion into contact with the absorptive surface of the intestinal cells. Then, **fatty acids, monoglycerides, and cholesterol diffuse across the luminal membrane into the cells.** Glycerol is hydrophilic and is not contained in the micelles.

  b. In the intestinal cells, the products of lipid digestion are **reesterified to triglycerides and phospholipids,** and, with cholesterol and apoproteins, form **chylomicrons**.
    −**Failure to synthesize apoprotein B** results in the inability to transport chylomicrons out of the intestinal cells and causes β-**lipoproteinemia**.

  c. **Chylomicrons are transported out of the intestinal cells by exocytosis.** Because chylomicrons are too large to enter the capillaries, they are transferred to **lymph vessels** and are added to the bloodstream via the **thoracic duct**.

3. Malabsorption of lipids—steatorrhea
  −can be caused by the following:

  a. **Pancreatic disease** (pancreatitis, cystic fibrosis)

  b. **Hypersecretion of gastrin,** which increases gastric H^+ secretion and lowers the duodenal pH. Low pH inactivates pancreatic lipase.

  c. **Deficiency of bile acids** from ileal resection, or inactivation of bile acids by bacterial overgrowth

  d. **Decreased number of intestinal cells** for absorption (tropical sprue)

  e. **Failure to synthesize apoprotein B** and inability to absorb chylomicrons

D. Absorption and secretion of electrolytes and H_2O
  −Electrolytes and H_2O may cross intestinal epithelial cells by either **cellular** or **paracellular** (between cells) routes.
  −**Tight junctions** attach the epithelial cells to one another at the luminal membrane. The permeability of the tight junctions varies with the type of epithelial cell.
  −A **"tight"** (impermeable) epithelium is the colon.
  −**"Leaky"** (permeable) epithelia are the small intestine and gallbladder.

1. Absorption of NaCl

  a. Na^+ moves into the intestinal cells, across the luminal membrane, and down its electrochemical gradient by the following mechanisms:
    (1) Passive diffusion (Na^+ channels)

    (2) Na^+–glucose or Na^+–amino acid cotransport

    (3) Na^+–Cl^- cotransport

    (4) Na^+–H^+ exchange

    –In the **small intestine,** the Na^+-cotransport and Na^+–H^+ exchange mechanisms are most important. These cotransport and exchange mechanisms are similar to those in the renal proximal tubule.

    –In the **colon,** passive diffusion via Na^+ channels is most important. The Na^+ channels of the colon are similar to those in the renal distal tubule and are stimulated by **aldosterone.**

  b. Na^+ is pumped out of the cell against its electrochemical gradient by the Na^+–K^+ pump on the basolateral membranes.

  c. Cl^- absorption accompanies Na^+ absorption throughout the GI tract by the following mechanisms:

    (1) Passive diffusion by a paracellular route

    (2) Na^+–Cl^- cotransport

    (3) Cl^-–HCO_3^- exchange

2. Absorption and secretion of K^+

  a. Dietary K^+ is **absorbed in the small intestine** by passive diffusion via a paracellular route.

  b. K^+ is actively **secreted by the colon** by a mechanism similar to that for K^+ secretion by the renal distal tubule. In fact, K^+ secretion by the colon is stimulated by **aldosterone.**

    –**In diarrhea, K^+ secretion by the colon is increased** because of a flow rate-dependent mechanism much like that of the renal distal tubule. The excess loss of K^+ in diarrheal fluid accounts, in part, for the observed **hypokalemia.**

3. Absorption of H_2O

  –is secondary to solute absorption.

  –is **isosmotic in the small intestine and gallbladder.** The mechanism for coupling solute and water absorption in these epithelia is the same as that described for the **renal proximal tubule.**

  –In the **colon,** the H_2O permeability is much lower than in the small intestine, and the feces may be hypertonic.

4. Secretion of electrolytes and H_2O by the intestine

  –In addition to the ability to absorb ingested electrolytes and H_2O, the GI tract can also secrete them from blood to lumen.

  –The **secretory mechanisms are located in the crypts,** whereas the absorptive mechanisms are in the villi.

  a. Cl^- **is the primary ion secreted** into the intestinal lumen. It moves through Cl^- channels that are regulated by **cyclic AMP.**

  b. Na^+ is secreted into the lumen by passively following Cl^-. H_2O follows to maintain isosmolarity.

  c. *Vibrio cholerae* (**cholera toxin**) causes diarrhea by activating adenylate cyclase in the basolateral membrane of the crypt cells, **increasing intracellular cyclic AMP,** and **activating the Cl^- secretory**

channels. Na^+ and H_2O follow Cl^- into the lumen. Certain strains of *Escherichia coli* cause diarrhea by a similar mechanism.

E. Absorption of other substances

1. Vitamins

a. Fat-soluble vitamins (A, D, E, and K) are included in micelles and absorbed along with other lipids.

b. Water-soluble vitamins are absorbed by Na^+–dependent cotransport mechanisms.

–**Vitamin B_{12} is absorbed in the ileum and requires intrinsic factor.** Vitamin B_{12}–intrinsic factor complex binds to a receptor on the ileal cells and is absorbed.

–**Gastrectomy** results in loss of gastric parietal cells as the source of intrinsic factor. Injection of vitamin B_{12} is required to prevent **pernicious anemia**.

2. Calcium

–absorption in the small intestine depends on adequate amounts of the active form of vitamin D, **1,25-dihydroxycholecalciferol**, which is produced in the kidney.

–Vitamin D deficiency or renal failure result in inadequate intestinal Ca^{2+} absorption, causing **rickets** (children) or **osteomalacia** (adults).

3. Iron

–is **absorbed as "heme iron"** (bound to hemoglobin or myoglobin) **or as free Fe^{2+}**. Once in the intestinal cell, "heme iron" is broken down; when free iron is released, it binds to apoferritin and eventually is transported into the blood.

–**circulates bound to transferrin**, which transports it from the small intestine to its storage sites in the liver, and from the liver to the bone marrow for synthesis of hemoglobin.

–Iron deficiency is the most common cause of anemia.

Review Test

Directions: Each of the numbered items or incomplete statements in this section is followed by answers or by completions of the statement. Select the **one** lettered answer or completion that is **best** in each case.

1. Slow waves in small intestinal smooth muscle cells are

(A) action potentials
(B) phasic contractions
(C) tonic contractions
(D) oscillating resting membrane potentials
(E) oscillating release of CCK

2. When parietal cells are stimulated, they secrete

(A) HCl and intrinsic factor
(B) HCl and pepsinogen
(C) HCl and HCO_3^-
(D) HCO_3^- and intrinsic factor
(E) mucus and pepsinogen

3. *Vibrio cholerae* causes diarrhea because it

(A) increases HCO_3^- secretory channels in intestinal epithelial cells
(B) increases Cl^- secretory channels in intestinal epithelial cells
(C) prevents absorption of glucose and causes water to be retained in the intestinal lumen isosmotically
(D) inhibits cyclic AMP production in intestinal epithelial cells
(E) inhibits IP_3 production in intestinal epithelial cells

4. CCK has some gastrin-like properties because both CCK and gastrin

(A) are released from G cells in the stomach
(B) are released from I cells in the duodenum
(C) are members of the secretin-homologous family
(D) have five identical carboxy-terminal amino acids
(E) have 90% homology of their amino acids

5. All of the following are transported in intestinal epithelial cells by a Na^+–dependent cotransport process EXCEPT

(A) glucose
(B) galactose
(C) fructose
(D) amino acids
(E) dipeptides

6. A patient with severe Crohn's disease has been unresponsive to drug therapy and undergoes ileal resection. Following the surgery, he will have steatorrhea because

(A) the liver bile acid pool increases
(B) chylomicrons fail to form in the intestinal lumen
(C) micelles fail to form in the intestinal lumen
(D) dietary triglycerides cannot be digested
(E) the pancreas does not secrete lipase

7. CCK stimulates all of the following EXCEPT

(A) gastric emptying
(B) pancreatic HCO_3^- secretion
(C) pancreatic enzyme secretion
(D) contraction of the gallbladder
(E) relaxation of the sphincter of Oddi

8. Which of the following would abolish "receptive relaxation" of the stomach?

(A) Parasympathetic stimulation
(B) Sympathetic stimulation
(C) Vagotomy
(D) Administration of gastrin

9. Peristalsis of the small intestine

(A) mixes the food bolus
(B) is coordinated by the central nervous system (CNS)
(C) involves contraction of smooth muscle behind and in front of the food bolus
(D) involves contraction of smooth muscle behind the food bolus and relaxation of smooth muscle in front of the bolus
(E) involves relaxation of smooth muscle throughout the small intestine

10. A patient with Zollinger-Ellison syndrome would be expected to have all of the following changes EXCEPT

(A) increased serum gastrin levels
(B) increased serum insulin levels
(C) increased gastric H^+ secretion
(D) increased parietal cell mass
(E) peptic ulcer disease

11. Micelle formation is necessary for the intestinal absorption of all of the following EXCEPT

(A) cholesterol
(B) fatty acids
(C) bile acids
(D) vitamin K
(E) vitamin D

12. Which of the following changes occurs during defecation?

(A) Internal anal sphincter is relaxed
(B) External anal sphincter is contracted
(C) Rectal smooth muscle is relaxed
(D) Intra-abdominal pressure is lower than when at rest
(E) Segmentation contractions predominate

13. Which of the following is characteristic of saliva?

(A) Hypotonicity relative to plasma
(B) A lower HCO_3^- concentration than plasma
(C) The presence of proteases
(D) Secretion rate increased by vagotomy
(E) Modification by the salivary ductal cells, which reabsorb K^+ and HCO_3^-

14. A patient with a duodenal ulcer is treated successfully with the drug cimetidine. The basis for cimetidine's inhibition of gastric H^+ secretion is that it

(A) blocks muscarinic receptors on parietal cells
(B) blocks histamine (H_2) receptors on parietal cells
(C) increases intracellular cyclic AMP levels
(D) blocks H^+–K^+ ATPase
(E) enhances the action of ACh on parietal cells

15. All of the following statements are true about secretion from the exocrine pancreas EXCEPT

(A) it has a higher Cl^- concentration than does plasma
(B) it is stimulated by the presence of fatty acids in the duodenum
(C) HCO_3^- secretion is increased by secretin
(D) enzyme secretion is increased by CCK
(E) it is isotonic

16. All of the following products of digestion are absorbed by specific carriers in intestinal cells EXCEPT

(A) fructose
(B) sucrose
(C) alanine
(D) dipeptides
(E) tripeptides

Directions: Each group of items in this section consists of lettered options followed by a set of numbered items. For each item, select the **one** lettered option that is most closely associated with it. Each lettered option may be selected once, more than once, or not at all.

Questions 17–20

Match each numbered characteristic below with the appropriate GI hormone.

(A) Secretin
(B) Gastrin
(C) CCK
(D) VIP
(E) GIP

17. Released from neurons in the GI tract and produces smooth muscle relaxation

18. Low pH inhibits its release

19. Low pH in the duodenal lumen stimulates its release

20. Secreted in response to an oral glucose load

Questions 21–23

Match each numbered phenomenon below with the correct portion of the GI tract.

(A) Gastric antrum
(B) Gastric fundus
(C) Duodenum
(D) Ileum
(E) Colon

21. Secretion of intrinsic factor

22. Secretion of K^+

23. Secretion of gastrin

Answers and Explanations

1–D. Slow waves are oscillating resting membrane potentials of the GI smooth muscle. The slow waves bring the membrane potential toward or to threshold, but *are not themselves action potentials*. If the membrane potential is brought to threshold by a slow wave, then action potentials occur followed by contraction.

2–A. The gastric parietal (oxyntic) cells secrete HCl and intrinsic factor. The chief cells (and mucous cells) secrete pepsinogen.

3–B. Cholera toxin activates adenylate cyclase (and increases cyclic AMP in the intestinal crypt cells). In the crypt cells, cyclic AMP activates the Cl^-–secretory channels and produces a primary secretion of Cl^-, with Na^+ and H_2O following.

4–D. The two hormones have five identical amino acids at the C terminus. Biologic activity of CCK is associated with the seven C-terminal amino acids and biologic activity of gastrin is associated with the four C-terminal amino acids. Since this CCK heptapeptide contains the five common amino acids, it is logical that CCK should have some gastrin-like properties.

5–C. Fructose is the only monosaccharide not absorbed by Na^+–dependent cotransport; it is transported by facilitated diffusion. Amino acids, dipeptides, and tripeptides (the products of protease digestion) are also absorbed by Na^+–dependent cotransporters.

6–C. Ileal resection removes the portion of the small intestine that normally transports bile acids from the lumen of the gut and recirculates them to the liver. Because this process normally maintains the bile acid pool, new synthesis of bile acids is needed only to replace those bile acids lost in the feces. With ileal resection, most of the bile acids secreted are excreted in the feces, and the liver pool is significantly diminished. Bile acids are needed for micelle formation in the intestinal lumen to solubilize the products of lipid digestion so that they can be absorbed. Chylomicrons are formed *within* the intestinal epithelial cells and are transported to lymph vessels.

7–A. CCK stimulates both functions of the exocrine pancreas: HCO_3^- secretion and digestive enzyme secretion. It also facilitates the delivery of bile from the gallbladder to the small intestinal lumen by causing contraction of the gallbladder while relaxing the sphincter of Oddi. CCK inhibits gastric emptying, which helps to slow the delivery of food from the stomach to the intestine during periods of high digestive activity.

8–C. Receptive relaxation of the orad region of the stomach is initiated when the stomach distends as food enters it from the esophagus. Because this is a parasympathetic (vagovagal) reflex, it would be abolished by vagotomy.

9–D. Peristalsis is contractile activity coordinated by the enteric nervous system (not the CNS) that serves to propel the intestinal contents forward. Normally, it occurs after mixing, digestion, and absorption have been sufficient. To propel the food bolus forward, the smooth muscle must contract behind it and simultaneously relax in front of it.

10–B. Zollinger-Ellison syndrome (gastrinoma) is a tumor of the non-beta cell pancreas. The tumor secretes gastrin, which then circulates to the gastric parietal cells to produce increased H^+ secretion, peptic ulcer, and increased parietal cell growth (trophic effect of gastrin). Because the tumor does not involve the pancreatic beta cells, insulin levels should be unaffected.

11–C. Micelles provide a mechanism for solubilizing fat-soluble nutrients in the aqueous solution of the intestinal lumen until they can be brought into contact with and absorbed by the intestinal epithelial cells. Cholesterol and fatty acids are products of lipid digestion in the intestinal lumen. Vitamins K and D are fat-soluble vitamins. Bile acids, while a key ingredient of micelles, are themselves absorbed by a specific Na^+–dependent cotransporter in the ileum.

12–A. During defecation, the internal and external anal sphincters must both be relaxed in order for feces to pass from the body. Rectal smooth muscle contracts, and intra-abdominal pressure is elevated by expiring against a closed glottis (Valsalva maneuver).

13–A. Saliva is characterized by hypotonicity, a high HCO_3^- concentration, and the presence of α-amylase and lingual lipase (not proteases). The high HCO_3^- concentration is achieved by secretion of HCO_3^- into saliva by the ductal cells (not reabsorption of HCO_3^-). Because control of saliva production is parasympathetic, vagotomy abolishes it.

14–B. Cimetidine is a reversible inhibitor of H_2 receptors on parietal cells. Because histamine stimulates H^+ secretion, cimetidine blocks H^+ secretion. Levels of cyclic AMP (the second messenger for histamine) levels would be expected to fall. Cimetidine also blocks the action of ACh to stimulate H^+ secretion. Omeprazole is a pharmacologic agent that blocks H^+–K^+ ATPase directly.

15–A. The major anion in pancreatic secretions is HCO_3^- (which is in higher concentration than in plasma), so the Cl^- concentration is lower than in plasma. Pancreatic secretion is stimulated by the presence of fatty acids in the duodenum; secretin stimulates pancreatic HCO_3^- secretion and CCK stimulates pancreatic enzyme secretion.

16–B. Carbohydrates must be hydrolyzed to monosaccharides in the intestinal lumen to be transported into the intestinal epithelial cells for absorption; oligosaccharides and disaccharides are not transported. On the other hand, proteins are hydrolyzed to amino acids, dipeptides, or tripeptides, and all three forms are transported into intestinal cells for absorption.

17–D. VIP is a GI neurocrine that causes relaxation of GI smooth muscle. For example, VIP mediates the relaxation response of the lower esophageal sphincter when a bolus of food approaches it, allowing passage of the bolus into the stomach.

18–B. Gastrin's principal physiologic action is to increase H^+ secretion. H^+ secretion lowers the pH of the stomach contents, which, in turn, inhibits further secretion of gastrin—a classic example of negative feedback.

19–A. Secretin is released when gastric contents are delivered to the duodenum—specifically, H^+ and fatty acids. Low pH in the duodenum causes secretion of secretin. In turn, secretin causes secretion of pancreatic fluid high in HCO_3^-, which neutralizes the H^+ in the duodenal lumen.

20–E. GIP is the only GI hormone released in response to all three categories of nutrients—fat, protein, and carbohydrate. Oral glucose releases GIP, which, in turn, causes release of insulin from the endocrine pancreas. This action of GIP explains why oral glucose is more effective than intravenous glucose in releasing insulin.

21–B. Intrinsic factor is secreted by the parietal cells of the gastric fundus (as is HCl).

22–E. K^+ is reabsorbed by the small intestine and then secreted by the colon via a mechanism similar to that in the renal distal tubule (stimulated by aldosterone; sensitive to flow rate).

23–A. Gastrin is secreted by the G cells of the gastric antrum. HCl and intrinsic factor are secreted by the fundus.

7

Endocrine Physiology

I. Overview of Hormones

A. See Table 7-1 for a list of hormones, including abbreviations, gland of origin, and major actions.

B. Radioimmunoassay

—**Hormone concentrations** can be measured by highly specific and sensitive radioimmunoassays, as follows:

1. A serum sample with an unknown concentration of hormone is mixed with a known amount of antibody to the hormone and a known amount of radioactively labeled hormone.

2. The radioactively labeled and unlabeled (unknown) hormone compete for a fixed number of antibody-binding sites. When there is more unlabeled hormone present in the serum sample (i.e., the higher the hormone concentration), there is less of the radioactive hormone that can be bound to the antibody (i.e., and the more radioactive hormone that will be free in solution).

3. **A standard curve is prepared.** Usually, the ratio of bound/free radioactive hormone is plotted versus unlabeled hormone concentration (Figure 7-1). The hormone concentration in the serum sample is read off the standard curve.

   a. The ratio of **bound/free radioactive hormone will be highest when no unlabeled hormone is present** to compete for antibody-binding sites.

   b. The ratio of **bound/free will be lowest when unlabeled hormone concentration is highest**.

C. Hormone synthesis

1. **Protein and peptide hormone synthesis**

   —Synthesis of **preprohormone** takes place on the **rough endoplasmic reticulum,** directed by a specific mRNA.

   —**Signal peptides are cleaved** from the preprohormone, resulting in **prohormone,** which is transported to the Golgi apparatus.

Table 7-1. Master List of Hormones

Hormone	Abbreviation	Gland of Origin	Major Actions*
Thyrotropin-releasing hormone	TRH	Hypothalamus	Stimulates secretion of TSH
Corticotropin-releasing hormone	CRH	Hypothalamus	Stimulates secretion of ACTH
Gonadotropin-releasing hormone	GnRH	Hypothalamus	Stimulates secretion of LH and FSH
Growth hormone-releasing hormone	GHRH	Hypothalamus	Stimulates secretion of growth hormone
Somatotropin release-inhibiting hormone (somatostatin)	SRIF	Hypothalamus	Inhibits secretion of growth hormone
Prolactin-inhibiting factor (dopamine)	PIF	Hypothalamus	Inhibits secretion of prolactin
Prolactin-stimulating factor		Hypothalamus	Stimulates secretion of prolactin
Thyroid-stimulating hormone (thyrotropin)	TSH	Anterior pituitary	Stimulates synthesis and secretion of thyroid hormones
Follicle-stimulating hormone	FSH	Anterior pituitary	Stimulates growth of follicles and estrogen secretion (ovary) Promotes sperm maturation (testes)
Luteinizing hormone	LH	Anterior pituitary	Stimulates ovulation, formation of corpus luteum, and estrogen and progesterone synthesis (ovary) Stimulates synthesis and secretion of testosterone (testes)
Growth hormone	GH	Anterior pituitary	Stimulates protein synthesis and overall growth
Prolactin		Anterior pituitary	Stimulates milk production and secretion
Adrenocorticotropic hormone	ACTH	Anterior pituitary	Stimulates synthesis and secretion of adrenal cortical hormones
β-lipotropin		Anterior pituitary	? in humans
Melanocyte-stimulating hormone	MSH	Anterior pituitary	Stimulates melanin synthesis (? humans)
Oxytocin		Posterior pituitary	Milk ejection; uterine contraction
Antidiuretic hormone (vasopressin)	ADH	Posterior pituitary	Stimulates H_2O reabsorption by renal collecting ducts
L-thyroxine Triiodothyronine	T_4 T_3	Thyroid gland	Skeletal growth; $\uparrow$ O_2 consumption; heat production; $\uparrow$ protein, fat, and carbohydrate utilization; maturation of nervous system (perinatal)
Glucocorticoids (cortisol)		Adrenal cortex	Stimulates gluconeogenesis; anti-inflammatory; immunosuppression
Estradiol		Ovary	Growth and development of female reproductive organs; proliferative phase of menstrual cycle
Progesterone		Ovary	Secretory phase of menstrual cycle
Testosterone		Testes	Spermatogenesis; male secondary sex characteristics
Parathyroid hormone	PTH	Parathyroid gland	$\uparrow$ serum $[Ca^{2+}]$
Calcitonin		Thyroid gland (parafollicular cells)	$\downarrow$ serum $[Ca^{2+}]$
Aldosterone		Adrenal cortex	$\uparrow$ renal Na^+ reabsorption $\uparrow$ renal K^+ secretion
1,25-dihydroxychole-calciferol		Kidney (activation site)	$\uparrow$ intestinal Ca^{2+} absorption $\uparrow$ bone mineralization
Insulin		Pancreas (beta cells)	$\downarrow$ blood [glucose]
Glucagon		Pancreas (alpha cells)	$\uparrow$ blood [glucose]
Human chorionic gonadotropin	HCG	Placenta	$\uparrow$ estrogen and progesterone synthesis
Human placental lactogen (human chorionic somatomammotropin)	HPL	Placenta	Actions like growth hormone and prolactin during pregnancy

*Refer to text for more complete description of each hormone.

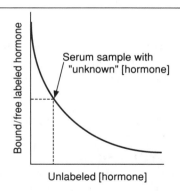

Figure 7-1. Using radioimmunoassay to measure hormone concentrations.

—Additional peptide sequences are cleaved in the Golgi apparatus to form the **hormone,** which is packaged in secretory granules for later release.

2. Steroid hormone synthesis

—Steroid hormones are **derivatives of cholesterol** (the biosynthetic pathways are described in V A 1).

3. Amine hormone synthesis

—Amine hormones (thyroid hormone, epinephrine, norepinephrine) are **derivatives of tyrosine**.

D. Regulation of hormone secretion

1. Negative feedback

—is the most commonly applied principle for regulating hormone secretion. It is self-limiting.

—**Hormones have biologic actions that,** directly or indirectly, **inhibit further secretion of the hormone**.

—**For example,** insulin is secreted by the pancreatic beta cells in response to an increase in blood glucose. In turn, insulin causes an increase in glucose utilization by cells, which results in decreased blood glucose concentration. The decrease in blood glucose concentration then tells the beta cells of the pancreas to decrease the secretion of insulin.

2. Positive feedback

—is explosive.

—A hormone has **biologic actions** that, directly or indirectly, **cause additional secretion of the hormone**.

—**For example,** the surge of luteinizing hormone (LH) just prior to ovulation is a result of positive feedback of estrogen on the anterior pituitary. LH then acts on the ovaries and causes more secretion of estrogen.

E. Regulation of receptors

—Hormones determine the sensitivity of the target tissue by **regulating the number or sensitivity of receptors**.

1. Down-regulation of receptors

—A hormone **decreases the number of receptors** for itself or for another hormone.

—**For example,** progesterone down-regulates its own receptor and the receptor for estrogen in the uterus.

2. Up-regulation of receptors
  –A hormone **increases the number of receptors** for itself or for another hormone.
  –**For example,** estrogen up-regulates its own receptor and the receptor for LH in the ovary.

II. Cell Mechanisms and Second Messengers (Table 7-2)

A. G proteins—general
  –are **guanosine triphosphate (GTP)-binding proteins** that couple hormone receptors to adjacent effector molecules. **For example,** in the cyclic adenosine monophosphate (AMP) second messenger system, G proteins couple the receptor to adenylate cyclase.
  –are used in the **adenylate cyclase, Ca^{2+}–calmodulin, and inositol triphosphate (IP_3) second messenger systems**.
  –have **intrinsic GTPase activity**.

B. Adenylate cyclase mechanism (Figure 7-2)

  1. Hormone binds to a receptor in the cell membrane (step 1). If the receptor is coupled to a stimulatory G protein (G_s), then adenylate cyclase will be activated. If the receptor is coupled to an inhibitory G protein (G_i), then adenylate cyclase will be inhibited.

  2. GDP is released from a binding site on the G protein, allowing GTP to bind to the G protein. It is the GTP plus G-protein complex that **activates adenylate cyclase** (step 2). Activated adenylate cyclase then catalyzes the conversion of ATP to cyclic AMP (step 3). Intrinsic GTPase activity converts GTP to guanosine diphosphate (GDP), restoring the original inactive state of the G protein.

  3. Cyclic AMP activates protein kinase A (step 4), which phosphorylates specific proteins (step 5), producing highly specific **physiologic actions** (step 6).

  4. Cyclic AMP is degraded to 5'-AMP by phosphodiesterase. Phosphodiesterase is inhibited by **caffeine** and **theophylline,** so these agents would be expected to augment the physiologic actions of cyclic AMP.

Table 7-2. Various Mechanisms of Hormone Action

Cyclic AMP Mechanism	IP_3 Mechanism	Steroid Hormone Mechanism	Other Mechanisms
ACTH	GnRH	Glucocorticoids	**Activation of tyrosine kinase**
LH and FSH	TRH	Estrogen	Insulin
TSH	GHRH	Testosterone	IGF-1
ADH (V_2 receptor)	Angiotensin II	Progesterone	
HCG	ADH (V_1 receptor)	Aldosterone	**Cyclic GMP**
MSH	Oxytocin	Vitamin D	ANP
CRH	α Receptors	Thyroid hormone	EDRF
β_1 and β_2 receptors			
Calcitonin			
PTH			
Glucagon			

Hormone

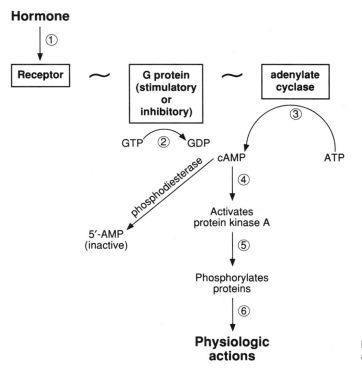

Figure 7-2. Mechanism of hormone action—adenylate cyclase.

C. **IP$_3$ mechanism** (Figure 7-3)

1. **Hormone binds to a receptor** in the cell membrane (step 1) and, via a G protein (step 2), **activates phospholipase C** (step 3).

2. **Phospholipase C liberates diacylglycerol and (IP$_3$)** from membrane bound lipids (step 4).

3. **IP$_3$ mobilizes Ca^{2+} from the endoplasmic reticulum** (step 5). Together, Ca^{2+} and diacylglycerol **activate protein kinase C** (step 6), which phosphorylates proteins and causes specific **physiologic actions** (step 7).

D. **Ca^{2+}–calmodulin mechanism** (Figure 7-4)

1. **Hormone binds to a receptor** in the cell membrane (step 1) and, via a G protein, produces an **increase in intracellular [Ca^{2+}]** (step 3) via two effects: opening cell membrane Ca^{2+} channels, and releasing Ca^{2+} from endoplasmic reticulum (step 2).

2. **Ca^{2+} binds to calmodulin** (step 4), and the Ca^{2+}–calmodulin complex produces **physiologic actions** (step 5).

E. **Steroid hormone and thyroid hormone mechanism** (Figure 7-5)

1. Steroid or thyroid hormones diffuse across the cell membrane (step 1) and bind to either a cytosolic or nuclear receptor. **Binding to the receptor** causes a conformational change in it, which **exposes a DNA-binding domain** (step 2).

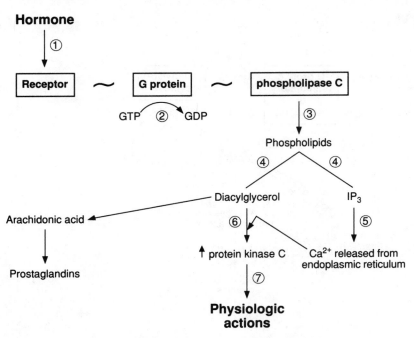

Hormone

① → **Receptor** ~ **G protein** ~ **phospholipase C**

GTP ② GDP

③

Phospholipids

④   ④

Diacylglycerol   IP$_3$

Arachidonic acid ◄

⑥   ⑤

Prostaglandins

↑ protein kinase C   Ca^{2+} released from endoplasmic reticulum

⑦

Physiologic actions

Figure 7-3. Mechanism of hormone action—IP$_3$ -Ca^{2+}.

2. In the nucleus, the DNA-binding domain on the receptor interacts with the hormone-regulatory elements of specific DNA. **Transcription is initiated** (step 3), resulting in production of mRNA.

3. **mRNA is translated** in the cytoplasm (step 4) and results in the **production of specific proteins** (steps 5 and 6) that have physiologic actions (e.g., the Ca^{2+}-binding protein induced by 1,25-dihydroxycholecalciferol).

III. Pituitary Gland (Hypophysis)

A. Hypothalamic–pituitary relationships

1. The **anterior lobe of the pituitary gland** is linked to the hypothalamus via the **hypothalamic–hypophysial portal system**. Thus, blood from the hypothalamus containing hypothalamic hormones is delivered directly and in high concentrations to the anterior pituitary.

2. The **posterior lobe of the pituitary gland** is derived from neural tissue. The **nerve cell bodies are located in hypothalamic nuclei**. Hormones of the posterior lobe of the pituitary are synthesized in the hypothalamus, packaged in secretory granules, and transported down the axons to be stored for release by the posterior lobe.

B. Hormones of the anterior lobe

–Growth hormone and prolactin are discussed in detail in this section. The other hormones of the anterior lobe are discussed in context (e.g., thyroid stimulating hormone [TSH] with thyroid hormone) in later sections of the chapter.

Hormone

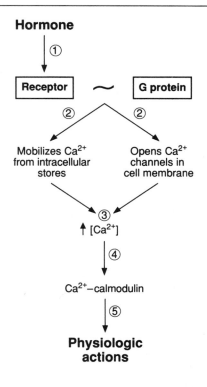

Figure 7-4. Mechanism of hormone action—Ca^{2+}-calmodulin.

1. **Thyroid-stimulating hormone (TSH), luteinizing hormone (LH), and follicle-stimulating hormone (FSH)**
   —are members of the same glycoprotein family, each having an α subunit and a β subunit. Because their **α subunits are identical**, the β subunits are responsible for the unique activity of each hormone.

2. **Adrenocorticotropic hormone (ACTH), melanocyte-stimulating hormone (MSH), β-lipotropin, and β-endorphin** (Figure 7-6)
   —are derived from a single precursor, **proopiomelanocortin (POMC)**.
   —**α-MSH and β-MSH** are produced in the intermediary lobe, which is rudimentary in adult humans.

3. **Growth hormone (somatotropin)**
   —is the most important hormone for normal growth to adult size.
   —is a single-chain polypeptide that is **homologous to human placental lactogen and prolactin**.
   a. **Regulation of secretion of growth hormone** (Figure 7-7)
      —Growth hormone is released in **pulsatile** fashion.
      —**Secretion is increased** by sleep, stress, hormones related to puberty, starvation, exercise, and hypoglycemia.
      —**Secretion is decreased** by somatostatin, somatomedins, obesity, hyperglycemia, and pregnancy.
      (1) **Hypothalamic control—growth hormone-releasing hormone (GHRH) and somatostatin**
         —**GHRH stimulates synthesis and release** of growth hormone.

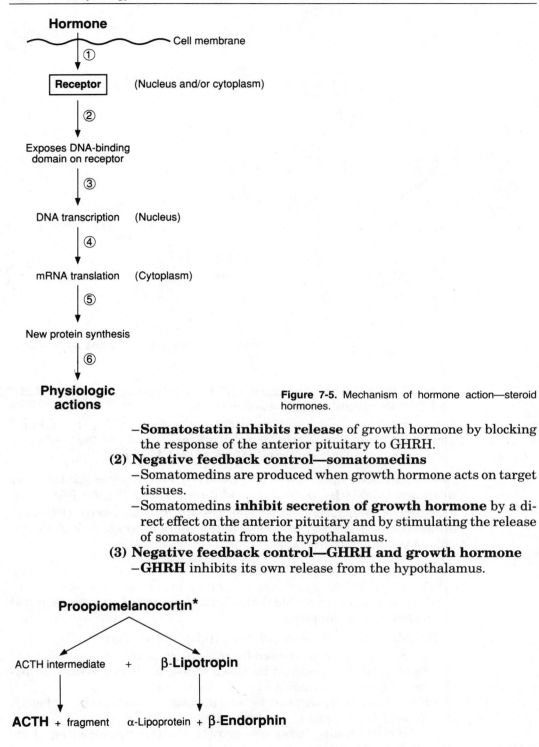

Figure 7-5. Mechanism of hormone action—steroid hormones.

  –**Somatostatin inhibits release** of growth hormone by blocking the response of the anterior pituitary to GHRH.

(2) Negative feedback control—somatomedins
  –Somatomedins are produced when growth hormone acts on target tissues.
  –Somatomedins **inhibit secretion of growth hormone** by a direct effect on the anterior pituitary and by stimulating the release of somatostatin from the hypothalamus.

(3) Negative feedback control—GHRH and growth hormone
  –**GHRH** inhibits its own release from the hypothalamus.

Figure 7-6. POMC is the precursor for ACTH, β-lipotropin, and β-endorphin in the anterior pituitary.

*Proopiomelanocortin = POMC

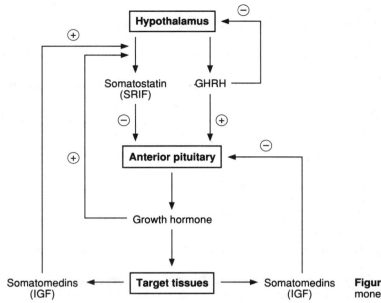

Figure 7-7. Control of growth hormone secretion.

—**Growth hormone** also inhibits its own secretion by stimulating somatostatin release from the hypothalamus.

b. **Actions of growth hormone**

—In the liver growth hormone generates the production of **somatomedins (insulin-like growth factor [IGF]),** which are the intermediaries of several physiologic actions of growth hormone.

—The **IGF receptor** has **tyrosine kinase activity,** similar to the insulin receptor.

(1) **Direct actions of growth hormone**

    (a) ↓ glucose uptake (**diabetogenic**)
    (b) ↑ lipolysis
    (c) ↑ protein synthesis in muscle and ↑ lean body mass
    (d) ↑ production of **IGF**

(2) **Actions of growth hormone via IGF**

    (a) ↑ protein synthesis in chondrocytes and ↑ **linear growth (pubertal growth spurt)**
    (b) ↑ protein synthesis in muscle and ↑ **lean body mass**
    (c) ↑ protein synthesis in most organs and ↑ **organ size**

c. **Pathophysiology of growth hormone**

(1) **Growth hormone deficiency**

    —in children, causes failure to grow, short stature, mild obesity, and delayed puberty.
    —can be caused by any of the following:
    (a) Lack of pituitary growth hormone
    (b) Hypothalamic dysfunction (↓ GHRH)
    (c) Failure to generate IGF in the liver
    (d) Receptor deficiency

(2) Growth hormone excess

—Hypersecretion of growth hormone causes **acromegaly**.

—**Somatostatin analogs** can be used for treatment of acromegaly because they inhibit growth hormone secretion by the anterior lobe of the pituitary.

(a) Before puberty, growth hormone excess causes increased linear growth (**gigantism**).

(b) After puberty, growth hormone excess causes increased periosteal bone growth, increased organ size, and glucose intolerance.

4. Prolactin

—is the major hormone responsible for **lactogenesis**.

—participates in breast development.

—is structurally **homologous to growth hormone**.

a. Regulation of prolactin secretion (Figure 7-8 and Table 7-3)

(1) Hypothalamic control—dopamine and thyrotropin-releasing hormone (TRH)

—Prolactin secretion is **tonically inhibited by dopamine** (prolactin-inhibiting factor [PIF]) from the hypothalamus. Thus, interruption of the hypothalamic–pituitary tract causes sustained lactation.

—**TRH increases prolactin secretion.**

(2) Negative feedback control

—Prolactin inhibits its own secretion by stimulating the hypothalamic release of dopamine.

b. Actions of prolactin

(1) ↑ **milk production** in the breast (casein, lactalbumin)

(2) ↓ synthesis and release of gonadotropin-releasing hormone (GnRH), which **inhibits ovulation**

(3) Inhibits spermatogenesis (via ↓ GnRH)

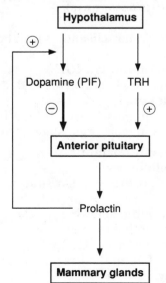

Figure 7-8. Control of prolactin secretion.

Table 7-3. Regulation of Prolactin Secretion

Produce Increased Prolactin Secretion	Produce Decreased Prolactin Secretion
Estrogen (pregnancy)	Dopamine
Breast feeding	Bromocriptine (dopamine agonist)
Sleep	Somatostatin
Stress	Prolactin (by negative feedback)
TRH	
Dopamine antagonists	

c. **Pathophysiology of prolactin**
  (1) **Prolactin deficiency (destruction of anterior pituitary)**
    —results in the **failure to lactate**.
  (2) **Prolactin excess**
    —causes **galactorrhea** and decreased libido.
    —causes **failure to ovulate** and amenorrhea because it inhibits GnRH secretion
    —**results from hypothalamic destruction** (due to loss of the normal "inhibitory" control by dopamine) or from prolactin-secreting tumors (**prolactinomas**).
    —can be treated with **bromocriptine,** which reduces serum prolactin levels by acting as a **dopamine agonist**.

C. **Hormones of the posterior lobe of the pituitary—antidiuretic hormone (ADH) and oxytocin**
—are homologous nonapeptides.
—are synthesized in hypothalamic nuclei and are packaged in secretory granules with their respective **neurophysins**.
—travel down the axon for secretion by the posterior pituitary.

1. **ADH** (see Chapter 5 VII for a detailed discussion)
  —originates primarily in the **supraoptic nuclei of the hypothalamus**.
  —regulates the serum osmolarity by increasing the H_2O permeability of the late distal tubules and collecting ducts.
  a. **Regulation of ADH secretion** (Table 7-4)
  b. **Actions of ADH**
    (1) ↑ **H_2O permeability** of late distal tubule and collecting duct (**V_2 receptor** with cyclic AMP mechanism)
    (2) **Constriction of vascular smooth muscle (V_1 receptor** with IP_3 mechanism)
  c. **Pathophysiology of ADH** (see Chapter 5 VI and VII)

Table 7-4. Regulation of ADH Secretion

Produce Increased ADH Secretion	Produce Decreased ADH Secretion
↑ serum osmolarity	↓ serum osmolarity
Volume contraction	Ethanol
Pain	α-Agonists
Nausea (powerful stimulant)	ANF
Hypoglycemia	
Nicotine, opiates, antineoplastic drugs	

2. Oxytocin
   —originates primarily in the **paraventricular nuclei of the hypothalamus**.
   —causes **milk ejection from the breast,** which is stimulated by suckling.

   a. Regulation of oxytocin secretion
   (1) Suckling
      —is the major stimulus for oxytocin secretion.
      —Afferent fibers carry impulses from the nipple to the spinal cord. Relays in the hypothalamus trigger oxytocin release from the posterior lobe.
      —The sight or sound of the infant may stimulate the hypothalamic neurons to secrete oxytocin.
   (2) Dilation of the cervix and orgasm
      —increase the secretion of oxytocin.

   b. Actions of oxytocin
   (1) Contraction of myoepithelial cells in the mammary glands
      —Milk is forced from the alveoli into the ducts and delivered to the infant.
   (2) Contraction of the uterus
      —During pregnancy, the number of oxytocin receptors in the uterus increases as parturition approaches, although the role of oxytocin in normal labor is uncertain.
      —Oxytocin can be used to induce labor and to **reduce postpartum bleeding**.

IV. Thyroid Gland

A. Synthesis of thyroid hormones (Figure 7-9)
   —Each step in synthesis is **stimulated by TSH**.

1. The iodide (I⁻) pump ("trap")
   —is present in the thyroid follicular epithelial cells.
   —actively transports I^- into the thyroid cells for subsequent incorporation into thyroid hormones.

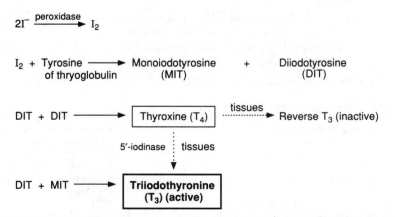

Figure 7-9. Steps in the synthesis of thyroid hormones. Each step is stimulated by TSH.

—is **inhibited by thiocyanate and perchlorate anions**.

—High doses of iodine (I_2) inhibit the synthesis of thyroid hormones (**Wolff-Chaikoff effect**).

2. Oxidation of I^- to I_2

—is catalyzed by a **peroxidase enzyme** in the follicular cell membrane. I_2 **is the reactive form,** which will be "organified" by combination with tyrosine.

—The oxidation reaction is **inhibited by propylthiouracil,** which is used therapeutically to reduce thyroid hormone synthesis in hyperthyroidism.

—The same peroxidase also catalyzes the organification and coupling reactions described below.

3. Organification of I_2

a. Thyroglobulin is synthesized on the ribosomes of the thyroid follicular cells, is packaged in secretory vesicles on the Golgi apparatus, and is then extruded into the follicular lumen.

b. At the junction of the follicular cells and the follicular lumen, tyrosine residues of thyroglobulin react with I_2 to form **monoiodotyrosine (MIT)** and **diiodotyrosine (DIT)**.

4. Coupling reaction

—While still part of thyroglobulin, two different coupling reactions involving MIT and DIT occur.

a. If two molecules of DIT combine, thyroxine (T_4) is formed.

b. If one molecule of DIT and one molecule of MIT combine, triiodothyronine (T_3) is formed.

5. Iodinated thyroglobulin

—is stored in the **follicular lumen** for later release of the thyroid hormones.

6. Stimulation of thyroid cells by TSH

—When the cells are stimulated, iodinated thyroglobulin must first be taken back into the follicular cells. Lysosomal enzymes then digest thyroglobulin, releasing T_4 and T_3 into the circulation.

—Leftover MIT and DIT are deiodinated by **thyroid deiodinase**. The I_2 that is released is reutilized for synthesis of more thyroid hormones. Therefore, **deficiency of thyroid deiodinase mimics I_2 deficiency**.

7. Binding of T_3 and T_4

—In the circulation, most of the T_3 and T_4 is bound to thyroxine-binding globulin (**TBG**).

8. Conversion of T_4 to T_3 and reverse T_3

—In the peripheral tissues T_4 is converted to T_3 or to reverse T_3.

—T_3 **is more biologically active than T_4,** and **reverse T_3 is inactive**. Thus, conversion of T_4 to T_3 is an activation step.

B. Regulation of thyroid hormone secretion (Figure 7-10)

1. Hypothalamic–pituitary control—TRH and TSH

a. TRH is secreted by the hypothalamus and stimulates secretion of TSH by the anterior lobe of the pituitary.

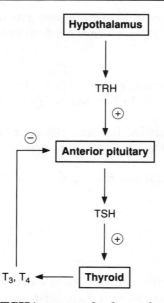

Figure 7-10. Control of thyroid hormone secretion.

    b. TSH increases both synthesis and secretion of thyroid hormones by the follicular cells via a **cyclic AMP mechanism**. Chronic elevation of TSH causes **hypertrophy** of the thyroid gland.

    c. Using negative feedback control, T_3 and T_4 inhibit the secretion of TSH from the anterior pituitary by decreasing the sensitivity of the secretory cells to TRH.

  2. Thyroid-stimulating immunoglobulins

    —are components of the IgG fraction of plasma proteins and are **antibodies to TSH receptors**.

    —react with the TSH receptor and, like TSH, **stimulate the thyroid gland to secrete T_3 and T_4**.

    —Circulate in large amounts in patients with **Graves' disease,** which is characterized by high circulating levels of thyroid hormones and accordingly low concentrations of TSH (caused by feedback inhibition of thyroid hormones on the anterior pituitary).

C. Actions of thyroid hormone

  —T_3 **is three to four times more potent than T_4**. Conveniently, the target tissues make T_3 from T_4 (see IV A 8).

  1. Growth

    —Attainment of adult stature requires thyroid hormone.

    —Thyroid hormones act synergistically with growth hormone and somatomedin to promote **bone formation**.

    —Thyroid hormones stimulate **bone maturation** as a result of ossification and fusion of the growth plates. **In thyroid hormone deficiency, bone age is less than chronologic age.**

  2. Central nervous system (CNS)

    a. Perinatal period

      —Maturation of the CNS is **absolutely dependent on thyroid hormone in the perinatal period.**

–Thyroid hormone deficiency causes irreversible mental retardation. Because there is only a brief perinatal period when thyroid hormone replacement therapy is helpful, **mandatory screening for neonatal hypothyroidism** has been instituted.

b. Adulthood

–**Hyperthyroidism** causes hyperexcitability and irritability.
–**Hypothyroidism** causes listlessness, slowed speech, somnolence, impaired memory, and decreased mental capacity.

3. Autonomic nervous system

–Thyroid hormone has many of the same actions as **β-adrenergic stimulation**. Therefore, a useful adjunct therapy for hyperthyroidism is treatment with a β-blocking agent, such as propranolol.

4. Basal metabolic rate (BMR)

–**O_2 consumption and BMR are increased by thyroid hormone** in all tissues **except brain, gonads, and spleen**. The resulting increase in heat production underlies the role of thyroid hormone in temperature regulation.
–Thyroid hormone **increases the synthesis of Na^+–K^+ ATPase,** and consequently increases O_2 consumption related to the Na^+–K^+ pump.

5. Cardiovascular and respiratory systems

–Effects of thyroid hormone on cardiac output and ventilation rate combine to ensure that more O_2 is delivered to the tissues.

a. ↑ heart rate and stroke volume → ↑ **cardiac output**

b. ↑ ventilation rate

6. Metabolic effects

–Overall, metabolism is increased to meet the demand for substrate associated with the increased rate of O_2 consumption.

a. ↑ glucose absorption from the gastrointestinal tract

b. ↑ **glycogenolysis, gluconeogenesis, and glucose oxidation** (driven by demand for ATP)

c. ↑ **lipolysis**

d. ↑ protein synthesis and degradation. The overall effect is **catabolic**.

D. Pathophysiology of the thyroid gland (Table 7-5)

V. Adrenal Cortex and Adrenal Medulla (Figure 7-11)

A. Adrenal cortex

1. Synthesis of adrenocortical hormones (Figure 7-12)

–The **zona fasciculata** produces mostly glucocorticoids (**cortisol**).
–The **zona reticularis** produces mostly androgens (**dehydroepiandrosterone and androstenedione**).
–The **zona glomerulosa** produces **aldosterone**.

a. 21-carbon steroids

–include **progesterone, deoxycorticosterone, aldosterone, and cortisol**.

Table 7-5. Pathophysiology of the Thyroid Gland

	Hyperthyroidism	**Hypothyroidism**
Symptoms	↑ metabolic rate Weight loss Negative nitrogen balance ↑ heat production (sweating) ↑ cardiac output Dyspnea Tremor, weakness Exophthalmos Goiter	↓ metabolic rate Weight gain Positive nitrogen balance ↓ heat production (cold sensitivity) ↓ cardiac output Hypoventilation Lethargy, mental slowness Drooping eyelids Myxedema Growth and mental retardation (perinatal) Goiter
Causes	Graves' disease (antibodies to TSH receptor) Thyroid neoplasm	Thyroiditis (autoimmune thyroiditis; Hashimoto's thyroiditis) Surgical destruction of thyroid I^- deficiency Cretinism (congenital) ↓ TRH or TSH
TSH levels	↓ (because of feedback inhibition on anterior pituitary by high thyroid hormone levels)	↑ (because of decreased feedback inhibition on anterior pituitary by low thyroid hormone levels) ↓ if primary defect is in hypothalamus or anterior pituitary
Treatment	Propylthiouracil (inhibits thyroid hormone synthesis by blocking oxidation of I^- to I_2) Thyroidectomy ^{131}I (destroys thyroid) β-blockers (adjunct therapy)	Thyroid hormone replacement

—Progesterone is the precursor for the others in the 21-carbon series.

—If the steroid nucleus is **hydroxylated at C-21,** deoxycorticosterone is produced, which has mineralocorticoid (but not glucocorticoid) activity.

—If the steroid nucleus is **hydroxylated at C-17,** then glucocorticoids (cortisol) are produced.

b. 19-carbon steroids

—have **androgenic activity** and are precursors to the estrogens.

—If the steroid has been previously hydroxylated at C-17, the $C_{20,21}$ side chain can be cleaved to yield the 19-carbon steroids **dehydroepiandrosterone or androstenedione** in the adrenal cortex.

—In the testes, androstenedione is converted to testosterone.

c. 18-carbon steroids

—have **estrogenic activity** and are compounds with an unsaturated A ring.

—Oxidation of the A ring (**aromatization**) occurs mainly in the **ovaries and placenta**.

2. Regulation of secretion

a. Glucocorticoid secretion (Figure 7-13)

—oscillates with a 24-hour periodicity or **circadian rhythm**.

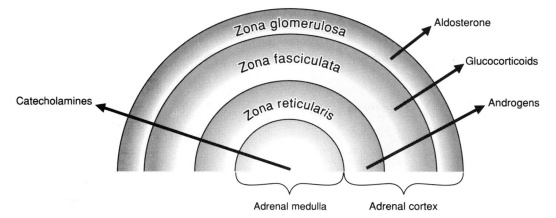

Figure 7-11. Secretory products of the adrenal cortex and medulla.

−For those who sleep during the night, **cortisol levels are highest just before waking** (≈ 6 A.M.) and **lowest in the evening** (≈ 12 midnight).

(1) Hypothalamic control—corticotropin-releasing factor (CRF)
  −CRF-containing neurons are located in the **paraventricular nuclei** of the hypothalamus.
  −When stimulated, CRF is released into hypophysial–portal blood.
  −CRF binds to receptors on the corticotropes of the anterior lobe of the pituitary and directs them to **synthesize POMC** (remember, it's the precursor to ACTH) and to **secrete ACTH**.
  −The second messenger for CRF is **cyclic AMP**.

(2) Anterior pituitary—ACTH
  −**ACTH stimulates steroidogenesis** by the zonae fasciculata and reticularis by **increasing the conversion of cholesterol to pregnenolone**.
  −ACTH also **up-regulates its own receptor** so that the sensitivity to subsequent doses of ACTH is increased.
  −Chronically increased levels of ACTH cause hypertrophy of the adrenal glands.
  −The second messenger for ACTH is **cyclic AMP**.

(3) Negative feedback control—cortisol
  −**Cortisol inhibits secretion of CRF** from the hypothalamus, and **inhibits secretion of ACTH** from the anterior pituitary.
  −If cortisol (glucocorticoid) levels are chronically elevated, then the synthesis of POMC (and therefore ACTH) is inhibited.
  −The **dexamethasone suppression test** is based on the ability of dexamethasone (a potent synthetic glucocorticoid) to inhibit ACTH and cortisol secretion. Suppression by negative feedback will occur if the hypothalamic–pituitary–adrenocortical axis is normal.

b. Aldosterone secretion (see Chapter 3 IV B)
  (1) Renin–angiotensin–aldosterone axis
    (a) Decreases in blood volume cause a decrease in renal perfusion pressure, which in turn increases renin secretion. **Renin**

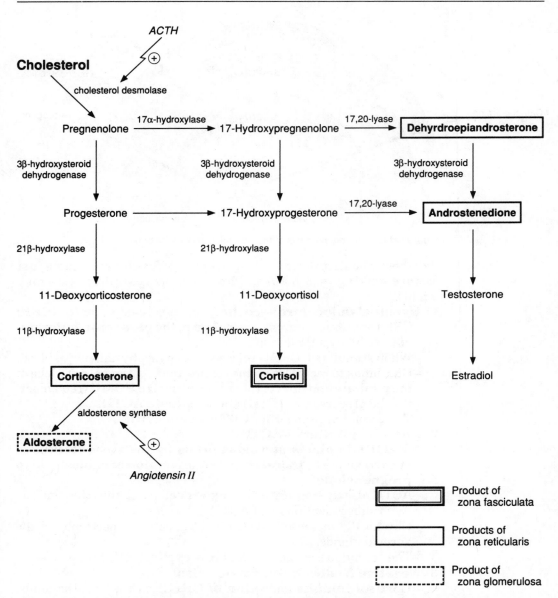

Figure 7-12. Synthetic pathways for glucocorticoids, androgens, and mineralocorticoids in the adrenal cortex.

catalyzes the conversion of angiotensinogen to angiotensin I. Angiotensin I is converted to **angiotensin II** by angiotensin-converting enzyme.

(b) **Angiotensin II** acts on the zona glomerulosa cells to **increase the conversion of corticosterone to aldosterone**.

(c) **Aldosterone** increases renal Na^+ reabsorption, thereby increasing extracellular fluid (ECF) volume and blood volume back to normal and inhibiting further renin secretion.

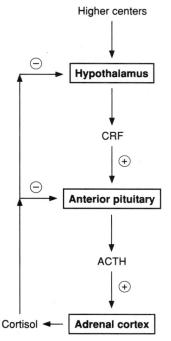

Higher centers

Hypothalamus

CRF

⊕

Anterior pituitary

ACTH

⊕

Cortisol ◄— Adrenal cortex

⊖ (two inhibitory signals shown on loop)

Figure 7-13. Control of glucocorticoid secretion.

(2) Hyperkalemia

—increases aldosterone secretion.

3. Actions of glucocorticoids (cortisol)

—Overall, glucocorticoids are essential for the **response to stress**.

a. Stimulation of gluconeogenesis

—Glucocorticoids increase gluconeogenesis by the following mechanisms:

(1) Increasing protein catabolism in muscle and decreasing protein synthesis, thereby providing more amino acids to the liver for gluconeogenesis.

(2) Decreasing glucose utilization and insulin sensitivity of adipose tissue.

(3) Increasing lipolysis, which provides more glycerol to the liver for gluconeogenesis.

b. Anti-inflammatory effects

(1) Glucocorticoids **induce the synthesis of lipocortin,** an **inhibitor of phospholipase A_2**. (Phospholipase A_2 is the enzyme that liberates arachidonate from membrane phospholipids, providing the precursor for prostaglandin and leukotriene synthesis.) Since prostaglandins and leukotrienes are involved in the inflammatory response, glucocorticoids have anti-inflammatory properties by inhibiting formation of the precursor (arachidonate).

(2) Glucocorticoids inhibit the production of interleukin-2 (IL-2) and inhibit proliferation of T lymphocytes.

(3) Glucocorticoids **inhibit the release of histamine and serotonin** from mast cells and platelets.

c. Suppression of the immune response

—Glucocorticoids (as noted above) **inhibit production of IL-2** and T lymphocytes, critical for cellular immunity. In pharmacologic doses, glucocorticoids are used to **prevent rejection of transplanted organs**.

d. Maintenance of vascular responsiveness to catecholamines

4. Actions of mineralocorticoids (aldosterone) [see Chapters 3 and 5]

    a. ↑ renal Na^+ reabsorption

    b. ↑ renal K^+ secretion

    c. ↑ renal H^+ secretion

5. Pathophysiology—adrenal cortex

    a. Adrenocortical insufficiency

      (1) Primary adrenocortical insufficiency—Addison's disease

        —is most commonly caused by autoimmune **destruction of the adrenal cortex** and causes acute **adrenal crisis**.

        —is characterized by the following:

        (a) ↓ **adrenal glucocorticoid, androgen, and mineralocorticoid**

        (b) ↑ **POMC and ACTH** (low cortisol levels decrease feedback inhibition on the anterior pituitary)

        (c) Hypoglycemia (because cortisol is hyperglycemic)

        (d) Weight loss, weakness, nausea, and vomiting

        (e) Hyperpigmentation

        (f) ↓ pubic and axillary hair in females (caused by absence of adrenal androgens)

        (g) ECF volume contraction, hyperkalemia, and **metabolic acidosis** (caused by absence of aldosterone)

      (2) Secondary adrenocortical insufficiency caused by decreased ACTH

        —is due to **primary deficiency of ACTH**.

        —**does not exhibit hyperpigmentation** (because there is a primary lack of ACTH).

        —**does not exhibit volume contraction, hyperkalemia, or metabolic acidosis** (because aldosterone is normal).

        —Symptoms are otherwise similar to Addison's disease.

    b. Adrenocortical excess—Cushing's syndrome

      —is most commonly caused by administration of **pharmacologic doses of glucocorticoids**.

      —is less commonly caused by bilateral **hyperplasia of the adrenal glands**.

      —is called **Cushing's disease** when it is caused by overproduction of ACTH.

      —is characterized by the following:

      (1) ↑ **cortisol and androgen levels**

(2) ↑ ACTH (only if caused by overproduction of ACTH)

(3) Hyperglycemia (caused by excess cortisol)

(4) ↑ protein catabolism and muscle wasting

(5) Central obesity (round face, supraclavicular fat, buffalo hump)

(6) Poor wound healing

(7) Virilization of the female (caused by excess adrenal androgens)

(8) Hypertension (caused by excess aldosterone)

(9) Osteoporosis (excess cortisol causes bone resorption)

(10) Striae

–**Ketoconazole**, an inhibitor of steroid hormone synthesis, can be used in the treatment of Cushing's disease.

c. Hyperaldosteronism—Conn's syndrome

–is caused by an aldosterone-secreting tumor.

–is characterized by the following:

(1) Hypertension (because aldosterone ↑ Na$^+$ reabsorption)

(2) Hypokalemia (because aldosterone ↑ K$^+$ secretion)

(3) Metabolic alkalosis (because aldosterone ↑ H$^+$ secretion)

(4) ↓ renin secretion (because of high levels of aldosterone)

d. Congenital abnormality—21β-hydroxylase deficiency

–is the most common biochemical abnormality of the steroidogenic pathway (see Figure 7-12). It belongs to a group of disorders characterized by **adrenogenital syndrome**.

–is characterized by the following:

(1) ↓ **cortisol and aldosterone** (resulting from enzyme block upstream)

(2) ↑ **17-hydroxyprogesterone** and progesterone (resulting from buildup behind the enzyme block)

(3) ↑ ACTH (resulting from decreased feedback inhibition by cortisol)

(4) Hyperplasia of zona fasciculata and zona reticularis (resulting from high levels of ACTH)

(5) ↑ **adrenal androgens** (because 17-hydroxyprogesterone is their major precursor)

(6) Virilization in females

(7) Early acceleration of linear growth and early appearance of pubic and axillary hair

(8) Suppression of gonadal function in males and females

B. Adrenal medulla (see Chapters 1 and 2)

VI. Endocrine Pancreas—Glucagon and Insulin (Table 7-6)

A. Organization of the endocrine pancreas

–The islets of Langerhans contain four cell types (Table 7-7).

–**Gap junctions** link beta cells to each other, alpha cells to each other, and beta cells to alpha cells for rapid communication.

–The portal blood supply of the islets allows blood from the beta cells to bathe the alpha and delta cells, again for rapid communication.

Table 7-6. Comparison of Insulin and Glucagon

	Stimulus for Secretion	Major Actions	Overall Effect on Blood Levels
Insulin (tyrosine kinase receptor)	↑ blood glucose ↑ amino acids ↑ fatty acids Glucagon GIP Growth hormone Cortisol	Increases glucose uptake into cells and glycogen formation Decreases glycogenolysis and gluconeogenesis Increases protein synthesis Increases fat deposition and decreases lipolysis Increases K^+ uptake into cells	↓ [glucose] ↓ [amino acid] ↓ [fatty acid] ↓ [ketoacid] Hypokalemia
Glucagon (cyclic AMP mechanism)	↓ blood glucose ↑ amino acids CCK Norepinephrine, epinephrine, ACh	Increases glycogenolysis and gluconeogenesis Increases lipolysis and keto-acid production	↑ [glucose] ↑ [fatty acid] ↑ [ketoacid]

B. Glucagon

1. Regulation of glucagon secretion (Table 7-8)

—The major factor regulating glucagon secretion is the blood glucose concentration. **Decreased blood glucose stimulates glucagon secretion.**

2. Actions of glucagon

—In contrast to insulin (which acts on the liver, adipose tissue, and muscle), glucagon **acts only on the liver**.

—The second messenger for glucagon is **cyclic AMP**.

a. Increases blood glucose concentration

(1) **Increases glycogenolysis** and prevents recycling of glucose into glycogen.

(2) **Increases gluconeogenesis.** Glucagon decreases the production of **fructose 2,6-bisphosphate,** decreasing **phosphofructokinase** activity; in effect, substrate is directed toward the formation of glucose rather than towards glucose breakdown.

b. Increases blood fatty acid and ketone concentration

—**Increases lipolysis.** Inhibition of fatty acid synthesis in effect "shunts" substrates towards gluconeogenesis.

—**Ketones (β-hydroxybutyrate and acetoacetate) are produced** from acetyl-coenzyme A (CoA), which results from fatty acid degradation.

Table 7-7. Cell Types of the Islets of Langerhans

Type of Cell	Location	Function
Beta	Central islet	Secrete insulin
Alpha	Outer rim of islet	Secrete glucagon
Delta	Intermixed	Secrete somatostatin and gastrin
Pancreatic polypeptide		Secrete pancreatic polypeptide

Table 7-8. Regulation of Glucagon Secretion

Cause Increased Glucagon Secretion	Cause Decreased Glucagon Secretion
↓ blood glucose	↑ blood glucose
↑ amino acids (especially arginine)	Insulin
CCK (alerts alpha cells to a protein meal)	Somatostatin
Norepinephrine, epinephrine	Fatty acids, ketones
ACh	

c. Increases urea production

–Amino acids are used for gluconeogenesis (stimulated by glucagon), and the resulting **amino groups will be incorporated into urea**.

C. Insulin

–is comprised of an A chain and a B chain, joined by two disulfide bridges.
–**Proinsulin is synthesized as a single chain.** The connecting peptide (C peptide) is removed by a protease within storage granules. Since the **connecting peptide** is secreted along with insulin, its concentration in blood is useful for monitoring beta cell function in diabetics receiving exogenous insulin.

1. Regulation of insulin secretion (Table 7-9)

a. Blood glucose concentration

–is the major factor regulating insulin secretion.
–**Increased blood glucose stimulates insulin secretion.** An initial burst of insulin is followed by sustained secretion.
–Glucose must be metabolized by the beta cells in order to stimulate insulin secretion.

b. Insulin secretion

–is mediated by an **opening of Ca^{2+} channels** in the beta cell membrane, an influx of Ca^{2+}, and a subsequent rise in intracellular $[Ca^{2+}]$.
–The influx of Ca^{2+} causes **depolarization of the beta cell,** similar to an action potential, and ultimately secretion of insulin.

2. Insulin receptor

–is a tetramer with two α subunits and two β subunits.

Table 7-9. Regulation of Insulin Secretion

Cause Increased Insulin Secretion	Cause Decreased Insulin Secretion
↑ blood glucose	↓ blood glucose
↑ amino acids (arginine, lysine, leucine)	Somatostatin
↑ fatty acids	Norepinephrine, epinephrine
Glucagon	(α receptors)
GIP	
ACh	
GH, cortisol	

a. The β subunits span the membrane and have tyrosine kinase activity. When insulin binds to the receptor, tyrosine kinase autophosphorylates the β subunits. The phosphorylated receptor then phosphorylates intracellular proteins.

b. The insulin-receptor complexes actually enter the target cells.

c. Insulin **down-regulates** its own receptors.
   – The number of insulin receptors is **increased in starvation,** and the number is **decreased in obesity**.

3. Actions of insulin

a. Decreases blood glucose concentration
   – by the following mechanisms:
   (1) Increases uptake of glucose by target cells by directing the insertion of glucose transporters into cell membranes. As glucose enters the cells, the blood glucose concentration is, accordingly, decreased.
   (2) Promotes the formation of glycogen from glucose in muscle and liver, and simultaneously **inhibits glycogenolysis**.
   (3) Decreases gluconeogenesis. Insulin increases the production of fructose 2,6-bisphosphate, increasing phosphofructokinase activity; in effect, substrate is directed away from the formation of glucose.

b. Decreases blood fatty acid and ketone concentrations
   – In adipose tissue, insulin **stimulates fat deposition** and **inhibits lipolysis**.
   – Ketone formation is inhibited in the liver because decreased fatty acid degradation provides less acetyl-CoA substrate for their formation.

c. Decreases blood amino acid concentration
   – Insulin stimulates amino acid uptake into cells, increases protein synthesis, and inhibits protein degradation. Thus, insulin is **anabolic**.

d. Decreases blood K^+ concentration
   – Insulin increases K^+ uptake into cells thereby decreasing blood $[K^+]$.

4. Insulin pathophysiology—diabetes mellitus
   – **Case study:** A woman is brought to the emergency room. She is hypotensive and breathing rapidly; her breath has the odor of ketones. Analysis of her blood shows severe hyperglycemia, hyperkalemia, and blood gases consistent with metabolic acidosis.
   – **Explanation:**

a. Hyperglycemia
   – is consistent with insulin deficiency.
   – In the absence of insulin, glucose uptake into cells is decreased, as is storage of glucose as glycogen.
   – If measured, the woman's blood would also have shown increased levels of amino acids (from increased protein catabolism) and increased levels of fatty acids (from increased lipolysis).

b. Hypotension
   – is a result of ECF volume contraction.

   −The high blood glucose concentration increases the filtered load of glucose to the point where it exceeds the reabsorptive ability (T_m) of the kidney.
   −The unreabsorbed glucose acts as an osmotic diuretic in the urine and causes ECF volume contraction.

   c. Metabolic acidosis
   −is a result of overproduction of ketoacids (β-hydroxybutyrate and acetoacetate).
   −The **increased ventilation rate** is the respiratory compensation for metabolic acidosis.

   d. Hyperkalemia
   −results from the lack of insulin, which normally promotes K^+ uptake into cells.
   −K^+ uptake into cells is decreased, so the blood $[K^+]$ increases.

D. Somatostatin
   −is secreted by the delta cells of the pancreas.
   −inhibits secretion of insulin, glucagon, and gastrin, and inhibits intestinal absorption of glucose.

VII. Calcium Metabolism (Parathyroid Hormone, Vitamin D, Calcitonin) [Table 7-10]

A. Overall Ca^{2+} homeostasis (Figure 7-14)
   −Serum $[Ca^{2+}]$ is determined by the interplay of **intestinal absorption, renal excretion,** and bone remodeling (**bone resorption and formation**). Each component is hormonally regulated.
   −To maintain Ca^{2+} balance, net intestinal absorption must be exactly balanced by urinary excretion.

1. Positive Ca^{2+} balance
   −is seen in growing children, where intestinal Ca^{2+} absorption exceeds urinary excretion, and the difference is deposited in the growing bones.

Table 7-10. Summary of Hormones Regulating Ca^{2+}

	PTH	Vitamin D	Calcitonin
Stimulus for secretion	↓ serum $[Ca^{2+}]$	↓ serum $[Ca^{2+}]$ ↑ PTH ↓ serum [phosphate]	↑ serum $[Ca^{2+}]$
Actions on:			
Bone	↑ resorption	↑ resorption	↓ resorption
Kidney	↓ P reabsorption (↑ urinary cyclic AMP) ↑ Ca^{2+} reabsorption	↑ P reabsorption ↑ Ca^{2+} reabsorption	
Intestine	↑ Ca^{2+} absorption (via vitamin D)	↑ Ca^{2+} absorption (vitamin D-dependent Ca^{2+}-binding protein) ↑ P absorption	
Overall effect on:			
Serum $[Ca^{2+}]$	↑	↑	↓
Serum [phosphate]	↓	↑	

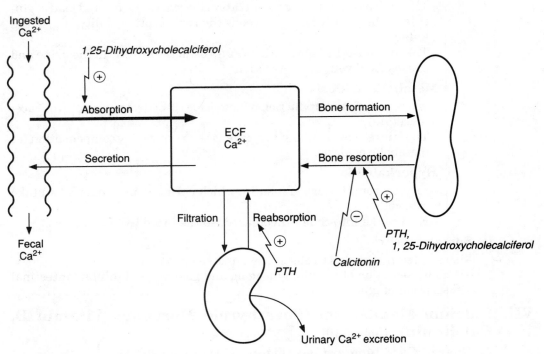

Figure 7-14. Hormonal regulation of Ca^{2+} metabolism.

2. Negative Ca^{2+} balance

–is seen in women during pregnancy or lactation, where intestinal Ca^{2+} absorption is less than urinary excretion and the difference comes from the maternal bones.

B. Parathyroid hormone (PTH)

–is the major hormone for regulation of the serum $[Ca^{2+}]$.
–is synthesized and secreted by the **chief cells** of the parathyroid glands.

1. Secretion of PTH

–is controlled by the serum $[Ca^{2+}]$ by negative feedback. **Decreased serum $[Ca^{2+}]$ increases PTH secretion.**
–Mild decreases in serum $[Mg^{2+}]$ also stimulate PTH secretion.
–Severe decreases in serum $[Mg^{2+}]$ inhibit PTH secretion and produce symptoms of hypoparathyroidism.
–The second messenger for PTH secretion by the parathyroid gland is **cyclic AMP.**

2. Actions of PTH

–are coordinated to produce an **increase in serum $[Ca^{2+}]$** and a **decrease in serum [phosphate].**
–The second messenger for PTH actions on its target tissues is **cyclic AMP.**

a. PTH increases bone resorption, which brings both Ca^{2+} and phosphate from bone mineral into the ECF. Alone, the effect on bone would not increase the serum $[Ca^{2+}]$ because phosphate complexes Ca^{2+}.

–Resorption of the organic matrix of bone is reflected in **increased hydroxyproline excretion.**

    b. PTH inhibits renal phosphate reabsorption in the **proximal tubule** and, therefore, increases phosphate excretion (**phosphaturic effect**). As a result, the phosphate resorbed from bone is excreted in the urine, allowing the serum $[Ca^{2+}]$ to increase.

      −Cyclic AMP generated as a result of the action of PTH on the proximal tubule is excreted in the urine (**nephrogenous cyclic AMP**).

    c. PTH increases renal Ca^{2+} reabsorption in the **distal tubule,** which also increases the serum $[Ca^{2+}]$.

    d. PTH increases intestinal Ca^{2+} absorption indirectly by stimulating the production of 1,25-dihydroxycholecalciferol (see VII C).

  3. Pathophysiology—PTH

    a. Primary hyperparathyroidism

      −is most commonly caused by **parathyroid adenoma.**

      −is characterized by the following:

        (1) ↑ serum $[Ca^{2+}]$

        (2) ↓ serum [phosphate]

        (3) ↑ urinary phosphate excretion (phosphaturic effect of PTH)

        (4) ↓ urinary Ca^{2+} excretion (caused by increased Ca^{2+} reabsorption)

        (5) ↑ urinary (nephrogenous) cyclic AMP

        (6) ↑ bone resorption

    b. Hypoparathyroidism

      −is most commonly the aftermath of **thyroid surgery** or is **congenital**.

      −is characterized by the following:

        (1) ↓ serum $[Ca^{2+}]$ and **tetany**

        (2) ↑ serum [phosphate]

        (3) ↓ urinary phosphate excretion

    c. Pseudohypoparathyroidism type Ia—Albright's hereditary osteodystrophy

      −is the result of **defective G protein** in kidney and bone, which causes end-organ **resistance to PTH.**

      −There is **hypocalcemia** and **hyperphosphatemia** (like hypoparathyroidism) that is not correctable by administration of exogenous PTH.

      −Circulating endogenous **PTH levels are elevated** (stimulated by hypocalcemia).

C. Vitamin D

  −provides Ca^{2+} and phosphate to extracellular fluid for normal bone mineralization.

  −In children, vitamin D deficiency causes **rickets;** in adults, vitamin D deficiency causes **osteomalacia.**

  1. Vitamin D metabolism (Figure 7-15)

    −The active form of vitamin D is **1,25-dihydroxycholecalciferol.** Its production in the kidney is catalyzed by 1α-hydroxylase.

    −**1α-hydroxylase activity is increased** by the following:

    a. ↓ serum $[Ca^{2+}]$

    b. ↑ PTH levels (secondary to ↓ Ca^{2+})

    c. ↓ serum [phosphate] (secondary to ↑ PTH)

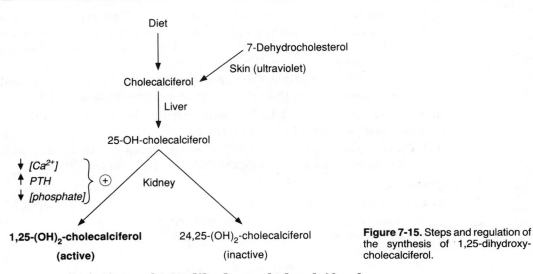

Diet

7-Dehydrocholesterol

Skin (ultraviolet)

Cholecalciferol

Liver

25-OH-cholecalciferol

↓ [Ca²⁺]
↑ PTH
↓ [phosphate]

(+) Kidney

1,25-(OH)₂-cholecalciferol

(active)

24,25-(OH)₂-cholecalciferol

(inactive)

Figure 7-15. Steps and regulation of the synthesis of 1,25-dihydroxy-cholecalciferol.

2. Actions of 1,25-dihydroxycholecalciferol

—are coordinated to **increase both Ca²⁺ and phosphate concentrations** in ECF in order to **mineralize new bone**.

a. Increases intestinal Ca²⁺ absorption. The protein induced by 1,25-dihydroxycholecalciferol is **vitamin D-dependent Ca²⁺-binding protein**.

—The action of PTH to increase intestinal Ca²⁺ absorption is indirect, via its stimulation of the 1α-hydroxylase and increased production of the active form of vitamin D.

b. Increases intestinal phosphate absorption.

c. Increases renal reabsorption of Ca²⁺ and phosphate, analogous to its actions on intestine.

d. Increases resorption of bone, which provides Ca²⁺ and phosphate from "old" bone to mineralize "new" bone.

D. Calcitonin

—is synthesized and secreted by the **parafollicular cells** of the thyroid.

—Secretion is stimulated by an increase in serum [Ca²⁺].

—Calcitonin's major action is to inhibit bone resorption.

VIII. Male Reproduction

A. Synthesis of testosterone (Figure 7-16)

—Testosterone is the major androgen synthesized and secreted by the **Leydig cells**.

—Leydig cells do not contain 21β-hydroxylase or 11β-hydroxylase (in contrast to the adrenal cortex) and do not synthesize glucocorticoids or mineralocorticoids.

—LH increases testosterone synthesis by stimulating cholesterol desmolase.

—**Accessory sex organs contain 5α-reductase,** which converts testosterone to dihydrotestosterone (the active form).

—**5α-reductase inhibitors (finasteride)** may be used in the treatment of **benign prostatic hypertrophy** because they block the activation of testosterone to dihydrotestosterone in the prostate.

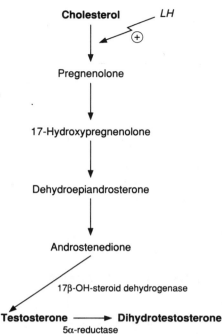

Cholesterol   LH

⊕

↓

Pregnenolone

↓

17-Hydroxypregnenolone

↓

Dehydroepiandrosterone

↓

Androstenedione

17β-OH-steroid dehydrogenase

Testosterone ──→ **Dihydrotestosterone**
5α-reductase
(target tissues)

Figure 7-16. Synthesis of testosterone.

B. Regulation of testes (Figure 7-17)

1. Hypothalamic control—GnRH

–Arcuate nuclei of the hypothalamus secrete GnRH into hypophysial–portal blood. GnRH stimulates the anterior pituitary to secrete FSH and LH.

2. Anterior pituitary—FSH and LH

–**FSH acts on the Sertoli cells** to maintain **spermatogenesis**. The Sertoli cells also secrete **inhibin,** which is involved in feedback inhibition of FSH secretion.

–**LH acts on the Leydig cells** to promote **testosterone synthesis**. Testosterone acts via an intratesticular paracrine mechanism to mediate the effects of FSH on spermatogenesis in the Sertoli cells.

3. Negative feedback control—testosterone and inhibin

–**Testosterone inhibits secretion of LH** by inhibiting release of GnRH from the hypothalamus and by directly inhibiting release of LH from the anterior pituitary.

–**Inhibin** (produced by Sertoli cells) **inhibits secretion of FSH** from the anterior pituitary.

C. Actions of testosterone

1. Causes prenatal differentiation of wolffian ducts and external genitalia (if 5α-reductase is present).

2. Causes development of male secondary sex characteristics at puberty (including male hair distribution, growth of external genitalia, laryngeal enlargement, increased muscle mass).

3. Causes pubertal growth spurt.

4. Maintains spermatogenesis in Sertoli cells (paracrine effect).

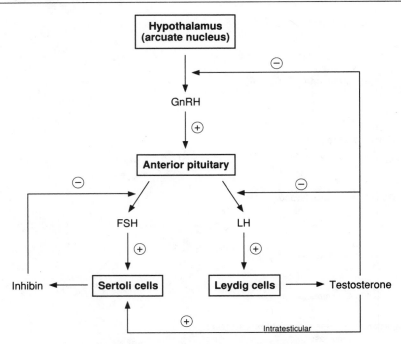

Figure 7-17. Control of male reproductive hormones.

5. Increases size and secretory activity of epididymis, vas deferens, prostate, and seminal vesicles (if 5α-reductase is present).

6. Increases libido.

D. Puberty (male and female)

—is initiated by the onset of **pulsatile GnRH release** from the hypothalamus.

—FSH and LH are, in turn, secreted in pulsatile fashion.

—GnRH **up-regulates** (and sensitizes) its receptor in the anterior pituitary.

IX. Female Reproduction

A. Synthesis of estrogen and progesterone (Figure 7-18)

—**Theca cells** produce mostly androgens (stimulated by LH). The androgens diffuse to the **granulosa cells,** which contain aromatase and convert androgen to estrogen (stimulated by FSH).

B. Regulation of the ovary

1. Hypothalamic control—GnRH

—As in the male, pulsatile GnRH stimulates the anterior pituitary to secrete FSH and LH.

2. Anterior pituitary—FSH and LH

—Stimulate the following:

a. Steroidogenesis in the ovarian follicle and corpus luteum

b. Follicular development beyond the antral stage

c. Ovulation

d. Luteinization

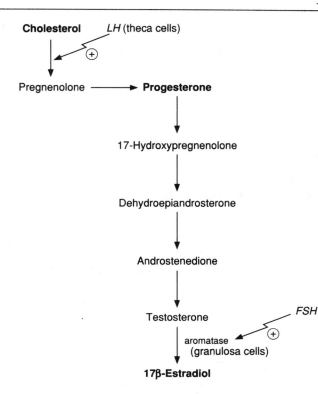

Figure 7-18. Synthesis of estrogen and progesterone.

3. **Negative and positive feedback control—estrogen and progesterone** (Table 7-11)

C. **Actions of estrogen**

1. Has both negative and positive feedback effects on FSH and LH secretion.

2. Causes maturation and maintenance of fallopian tubes, uterus, cervix, and vagina.

3. Is responsible for development of female secondary sex characteristics at puberty.

4. Is responsible for development of the breasts.

5. Up-regulates estrogen, LH, and progesterone receptors.

6. Is responsible for proliferation and development of granulosa cells.

7. Maintains pregnancy.

8. Lowers uterine threshold to contractile stimuli during pregnancy.

9. Stimulates prolactin secretion (but then blocks its action on the breast).

Table 7-11. Negative and Positive Feedback Control of the Menstrual Cycle

Phase of Menstrual Cycle	Hormone	Type of Feedback and Site
Follicular	Estrogen	Negative; anterior pituitary
Mid-cycle	Estrogen	Positive; anterior pituitary
Luteal	Estrogen	Negative; anterior pituitary
	Progesterone	Negative; hypothalamus

D. Actions of progesterone

1. Has negative feedback effects on FSH and LH secretion.
2. Maintains secretory activity of the uterus during the luteal phase.
3. Maintains pregnancy.
4. Raises uterine threshold to contractile stimuli during pregnancy.
5. Is responsible for development of the breasts.

E. Menstrual cycle (Figure 7-19)

1. **Follicular phase (days 5–14)**

   —**A primordial follicle develops** to the graafian stage with atresia of the neighboring follicles.
   —LH and FSH receptors are induced in theca and granulosa cells so that they can stimulate synthesis of androgen (LH, theca cells) and estradiol (FSH, granulosa cells).
   —**Estradiol levels steadily increase** and cause **proliferation of the uterus**.
   —**FSH and LH levels are suppressed** by negative feedback of estradiol on the anterior pituitary.
   —Progesterone levels are low.

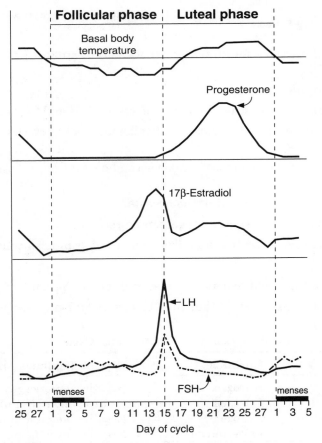

Figure 7-19. The menstrual cycle.

2. Ovulation (day 15)

–occurs 14 days prior to menses, regardless of cycle length. Thus, in a 35-day cycle, ovulation occurs on day 21.

–A burst of estradiol synthesis at the end of the follicular phase has a **positive feedback** effect on secretion of FSH and LH (**LH surge**).

–**Ovulation** occurs as a result of the **estrogen-induced LH surge**.

–Estrogen levels fall just after ovulation (but rise again in the luteal phase).

–**Cervical mucus increases** in quantity; it becomes less viscous and more penetrable by sperm.

3. Luteal phase (days 15–28)

–Development of the **corpus luteum** begins, and it **synthesizes estrogen and progesterone**.

–There is **increased vascularity and secretory activity of the endometrium** to prepare for receipt of a fertilized egg.

–**Basal body temperature increases** because of progesterone's effect on the hypothalamic thermoregulatory center.

–**If fertilization does not occur, the corpus luteum regresses.** As a result, there is an **abrupt decrease in estradiol and progesterone levels**.

4. Menses (days 1–4)

–The **endometrium is sloughed** because of the withdrawal of estradiol and progesterone.

F. Pregnancy (Figure 7-20)

–is characterized by steadily increasing levels of estrogen and progesterone, which maintain the endometrium for the fetus, suppress ovarian follicular function (by inhibiting FSH and LH secretion), and stimulate development of the breasts.

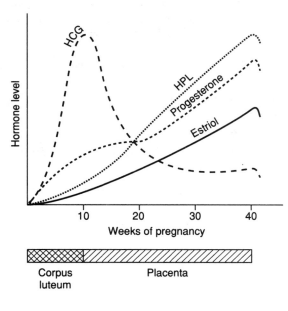

Figure 7-20. Hormone levels during pregnancy.

1. Fertilization
- If **fertilization occurs,** the corpus luteum is rescued from regression by **human chorionic gonadotrophin (HCG),** which is produced by the placenta.

2. First trimester
- The corpus luteum (stimulated by HCG) is responsible for production of estradiol and progesterone.
- Peak levels of HCG occur at gestational week nine and then decline.

3. Second and third trimesters
- **Progesterone** is produced by the placenta.
- **Estrogens** are produced by the interplay of the fetal adrenal gland and the placenta. The fetal adrenal gland synthesizes dehydroepiandrosterone-sulfate, which is then hydroxylated in the fetal liver; both are transferred to the placenta where enzymes remove sulfate and aromatize to estrogens. **The major placental estrogen is estriol.**
- **Human placental lactogen,** which has growth hormone-like and prolactin-like effects, is produced throughout pregnancy.

4. Parturition
- Throughout pregnancy, progesterone raises the threshold for uterine contraction.
- The **initiating event in parturition is unknown**. (Although oxytocin is a powerful stimulant of uterine contractions, there are no changes in blood levels of oxytocin prior to labor.)

5. Lactation
- Estrogens and progesterone stimulate growth and development of the breasts throughout pregnancy.
- **Prolactin increases steadily during pregnancy** because estrogen stimulates its secretion from the anterior pituitary.
- **Lactation does not occur during pregnancy because estrogen and progesterone block the action of prolactin on the breast.**
- After parturition, estrogen and progesterone levels fall rapidly so lactation can occur.
- Lactation is maintained by suckling, which stimulates both oxytocin and prolactin secretion.
- **Ovulation is suppressed** as long as lactation continues because prolactin has the following effects:

  a. Inhibits hypothalamic GnRH secretion.

  b. Inhibits the action of GnRH on the anterior pituitary, and consequently inhibits LH and FSH secretion.

  c. Antagonizes the actions of LH and FSH on the ovaries.

Review Test

Directions: Each of the numbered items or incomplete statements in this section is followed by answers or by completions of the statement. Select the **one** lettered answer or completion that is **best** in each case.

Questions 1–5

Use the graph below showing changes during the menstrual cycle to answer Questions 1–5.

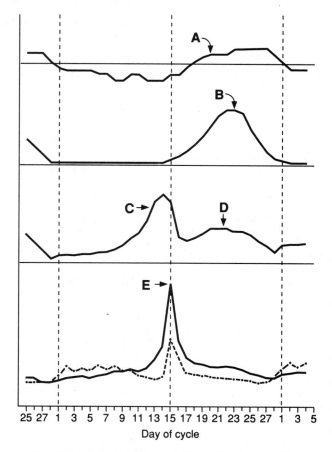

1. The increase shown at point **A** is caused by the effect of

(A) estrogen on the anterior pituitary
(B) progesterone on the hypothalamus
(C) FSH on the ovary
(D) LH on the anterior pituitary
(E) prolactin on the ovary

2. Blood levels of which substance are described by the curve labeled **B**?

(A) Estradiol
(B) Estriol
(C) Progesterone
(D) FSH
(E) LH

3. The source of the increase in concentration indicated at point **C** is the

(A) hypothalamus
(B) anterior pituitary
(C) corpus luteum
(D) ovary
(E) adrenal cortex

4. The source of the increase in concentration at point **D** is the

(A) ovary
(B) adrenal cortex
(C) corpus luteum
(D) hypothalamus
(E) anterior pituitary

5. The cause of the sudden increase shown at point **E** is

(A) negative feedback of progesterone on the hypothalamus
(B) negative feedback of estrogen on the anterior pituitary
(C) negative feedback of FSH on the ovary
(D) positive feedback of FSH on the ovary
(E) positive feedback of estrogen on the anterior pituitary

6. A woman has hypocalcemia, hyperphosphatemia, and decreased urinary phosphate excretion. Injection of PTH causes an increase in urinary cyclic AMP. The most likely diagnosis is

(A) primary hyperparathyroidism
(B) vitamin D intoxication
(C) vitamin D deficiency
(D) hypoparathyroidism following thyroid surgery
(E) pseudohypoparathyroidism

7. All of the following hormones act on their target tissues by an adenylate cyclase/cyclic AMP mechanism EXCEPT

(A) thyroid hormone
(B) parathyroid hormone
(C) ADH on the collecting duct
(D) β_1 adrenergic agonists
(E) glucagon

8. A man with galactorrhea is found to have a prolactinoma. His physician treats him with bromocriptine, which eliminates the galactorrhea. The basis for the therapeutic action of bromocriptine is that it

(A) antagonizes the action of prolactin on the breast
(B) enhances the action of prolactin on the breast
(C) inhibits prolactin release from the anterior pituitary
(D) inhibits prolactin release from the hypothalamus
(E) enhances the action of dopamine on the anterior pituitary

9. All of the following hormones are of hypothalamic origin EXCEPT

(A) dopamine
(B) GHRH
(C) somatostatin
(D) GnRH
(E) TSH

10. Which of the following functions of the Sertoli cells mediates negative feedback control of FSH secretion?

(A) Synthesis of inhibin
(B) Synthesis of testosterone
(C) Aromatization of testosterone
(D) Maintenance of the blood–testes barrier

11. Which of the following is NOT derived from proopiomelanocortin (POMC)?

(A) ACTH
(B) FSH
(C) β-lipotropin
(D) α-lipotropin
(E) β-endorphin

12. All of the following inhibit secretion of growth hormone by the anterior lobe of the pituitary gland EXCEPT

(A) sleep
(B) somatostatin
(C) growth hormone
(D) somatomedins
(E) obesity

13. All of the following are produced in the zona fasciculata and/or zona reticularis of the adrenal cortex EXCEPT

(A) aldosterone
(B) corticosterone
(C) cortisol
(D) dehydroepiandrosterone
(E) progesterone

14. Which of the following explains the suppression of lactation during the third trimester of pregnancy?

(A) Blood prolactin levels are too low
(B) Human placental lactogen levels are too low
(C) The fetal adrenal gland does not produce enough dehydroepiandrosterone
(D) Blood levels of estrogen and progesterone are high
(E) The maternal anterior pituitary is suppressed

15. Each of the following results from the action of PTH on the renal tubule EXCEPT

(A) stimulation of 1α-hydroxylase
(B) stimulation of Ca^{2+} reabsorption in the proximal tubule
(C) inhibition of phosphate reabsorption in the proximal tubule
(D) interaction with receptors on the basolateral membrane of proximal tubular cells
(E) increased urinary excretion of cyclic AMP

16. A female patient has hirsutism, hyperglycemia, obesity, muscle wasting, and increased circulating levels of ACTH. The most likely cause of her symptoms is

(A) primary adrenocortical insufficiency (Addison's disease)
(B) pheochromocytoma
(C) primary overproduction of ACTH (Cushing's disease)
(D) treatment with exogenous glucocorticoids
(E) hypophysectomy

17. All of the following increase the conversion of 25-hydroxycholecalciferol to 1,25-dihydroxycholecalciferol EXCEPT

(A) a diet low in Ca^{2+}
(B) hypocalcemia
(C) hyperparathyroidism
(D) hypophosphatemia
(E) chronic renal failure

18. Increased ACTH secretion would be expected in patients

(A) with chronic adrenocortical insufficiency (Addison's disease)
(B) with primary adrenocortical hyperplasia
(C) receiving glucocorticoid for immunosuppression following a renal transplant
(D) with elevated levels of angiotensin II

19. All of the following would be expected in a patient with Graves' disease EXCEPT

(A) increased sensitivity to cold temperatures
(B) weight loss
(C) increased O_2 consumption
(D) increased cardiac output
(E) increased ventilation rate

20. All of the following increase secretion of oxytocin EXCEPT

(A) suckling
(B) dilation of the cervix
(C) sight of the infant
(D) orgasm
(E) increased serum osmolarity

21. All of the following hormones act by a steroid hormone mechanism and induce synthesis of new protein EXCEPT

(A) 1,25-dihydroxycholecalciferol
(B) progesterone
(C) estradiol
(D) aldosterone
(E) GnRH

22. An untreated diabetic patient is brought to the emergency room. She would be expected to show all of the following EXCEPT

(A) glycosuria
(B) metabolic alkalosis
(C) hyperglycemia
(D) hyperkalemia
(E) hyperventilation

23. During the second and third trimesters of pregnancy, the source of elevated estrogen levels is the

(A) corpus luteum
(B) maternal ovaries and fetal adrenal gland
(C) maternal adrenal gland and fetal liver
(D) fetal adrenal gland and the placenta

24. Which of the following causes increased aldosterone secretion?

(A) Decreased blood volume
(B) Administration of an inhibitor of angiotensin-converting enzyme
(C) Hyperosmolarity
(D) Hypokalemia

Directions: Each group of items in this section consists of lettered options followed by a set of numbered items. For each item, select the **one** lettered option that is most closely associated with it. Each lettered option may be selected once, more than once, or not at all.

Questions 25–27

Match each numbered description involving thyroid hormones with the correct substance.

(A) T_3
(B) T_4
(C) DIT
(D) TSH
(E) I^-

25. Deiodination of this substance in the target tissues produces the most active form of thyroid hormone

26. In Graves' disease, blood levels of this hormone will be decreased

27. Propylthiouracil can be used to reduce synthesis of thyroid hormones in hyperthyroidism because it inhibits oxidation of this substance.

Questions 28–31

Match each numbered description with the correct step in steroid hormone biosynthesis.

(A) Cholesterol → pregnenolone
(B) Progesterone → 11-deoxycorticosterone
(C) 17-hydroxypregnenolone → dehydroepiandrosterone
(D) Testosterone → estradiol
(E) Testosterone → dihydrotestosterone

28. Step in steroid hormone biosynthesis that is stimulated by ACTH

29. If inhibited, blocks production of all androgenic compounds but does not block production of glucocorticoids

30. Is catalyzed by aromatase

31. Occurs in the accessory sex target tissues for testosterone and is catalyzed by 5α-reductase

Questions 32–34

Match each numbered description with the appropriate pancreatic secretory product.

(A) Insulin
(B) Glucagon
(C) Somatostatin
(D) Pancreatic lipase

32. Inhibits the secretion of both insulin and glucagon

33. Receptor has four subunits, two of which have tyrosine kinase activity

34. Is NOT secreted by the islets of Langerhans

Answers and Explanations

1–B. The curve labeled A is for basal body temperature. The rise in temperature occurs as a result of elevated progesterone levels during the luteal (secretory) phase of the menstrual cycle. Progesterone increases the set-point temperature in the thermoregulatory center in the hypothalamus.

2–C. Progesterone is secreted in the luteal phase of the menstrual cycle.

3–D. The curve is for blood levels of estradiol. The source of the rise in estradiol concentration shown at point C is the ovarian granulosa cells, which contain high concentrations of aromatase and convert testosterone to estradiol.

4–C. The curve is for blood levels of estradiol. In the luteal phase of the cycle, the source of the estradiol is the corpus luteum, not the ovarian follicle. The corpus luteum readies the uterus to receive a fertilized egg.

5–E. The point labeled E is the LH surge that initiates ovulation at mid-cycle. The LH surge is caused by increasing estrogen levels from the developing ovarian follicle, which, by positive feedback, stimulate the anterior pituitary to secrete LH and FSH.

6–D. Low blood $[Ca^{2+}]$ and high blood [phosphate] are consistent with some form of hypoparathyroidism. Lack of PTH would decrease bone resorption, decrease renal reabsorption of Ca^{2+}, and increase renal reabsorption of phosphate (low urinary phosphate). Since the patient responded to exogenous PTH with an increase in urinary cyclic AMP, the G protein that couples the PTH receptor to adenylate cyclase must be normal. Consequently, pseudohypoparathyroidism is ruled out.

7–A. Thyroid hormone, an amine, acts on its target tissues by a steroid-hormone type of mechanism, inducing new protein synthesis. The action of ADH on the collecting duct (V_2 receptors) is cyclic AMP-mediated, although the other action of ADH (vascular smooth muscle, V_1 receptors) is IP_3-mediated.

8–C. Bromocriptine is a dopamine agonist. Prolactin secretion by the anterior pituitary is tonically inhibited by secretion of dopamine from the hypothalamus. Thus, a dopamine agonist acts just like dopamine—it inhibits prolactin secretion from the anterior pituitary.

9–E. TSH is secreted by the anterior pituitary.

10–A. Inhibin is produced by the Sertoli cells of the testes when stimulated by FSH. In turn, inhibin inhibits further secretion of FSH by negative feedback on the anterior pituitary. The Leydig cells synthesize testosterone.

11–B. POMC is the parent molecule in the anterior pituitary for ACTH, β-endorphin, α-lipotropin, and β-lipotropin (and in the intermediary lobe, for MSH). FSH is not a member of this "family"; rather, it is related to TSH and LH.

12–A. Growth hormone is secreted in pulsatile fashion, with a large burst during deep sleep (sleep stage 3 or 4). Somatostatin is released from the hypothalamus and inhibits growth hormone secretion by the anterior pituitary. Growth hormone inhibits its own secretion (negative feedback by stimulating somatostatin release). Somatomedins are generated when growth hormone acts on its target tissues; they inhibit growth hormone secretion by the anterior lobe directly or indirectly by stimulating somatostatin release.

13–A. Aldosterone is not produced in the zona fasciculata and zona reticularis because those layers of the adrenal cortex do not contain the enzyme for conversion of corticosterone to aldosterone (aldosterone synthase). Aldosterone synthesis proceeds in the zona glomerulosa.

14–D. Although the high circulating levels of estrogen stimulate prolactin secretion during pregnancy, the action of prolactin on the breast is inhibited by progesterone and estrogen. Once parturition occurs, progesterone and estrogen levels fall dramatically. Prolactin can then interact with its receptors in the breast, and lactation proceeds if initiated by suckling.

15–B. PTH does stimulate renal Ca^{2+} reabsorption, but this action occurs in the distal tubule. It is the PTH inhibition of phosphate reabsorption that occurs in the proximal tubule.

16–C. This woman's symptoms are classic for a primary elevation of ACTH, which then stimulates overproduction of glucocorticoids and androgens. Treatment with pharmacologic doses of glucocorticoids would produce similar symptoms, except that circulating ACTH would be low because of negative feedback suppression at both hypothalamic (CRF) and anterior pituitary levels (ACTH). Addison's disease is caused by primary adrenocortical insufficiency. Although Addison's disease would have increased levels of ACTH (due to loss of negative feedback inhibition), the symptoms would be of glucocorticoid deficit, not excess. Hypophysectomy would remove the source of ACTH.

17–E. States of Ca^{2+} deficiency (low Ca^{2+} diet or hypocalcemia) activate the 1α-hydroxylase, which catalyzes the conversion of vitamin D to its active form, 1,25-dihydroxycholecalciferol. Increased PTH and hypophosphatemia also stimulate the enzyme. Chronic renal failure is associated with a constellation of bone diseases, including osteomalacia from failure of the diseased renal tissue to produce the active form of vitamin D.

18–A. Addison's disease is caused by primary adrenocortical insufficiency. The resulting decrease in cortisol production causes a decrease in negative feedback inhibition on the hypothalamus and the anterior pituitary, both of which will result in increased ACTH secretion. Patients with adrenocortical hyperplasia or those receiving exogenous glucocorticoid will have an increase in negative feedback inhibition of ACTH secretion.

19–A. Graves' disease (hyperthyroidism) is caused by overstimulation of the thyroid gland by circulating antibodies to the TSH receptor (which then increases production of thyroid hormone just as TSH would). T_3 increases O_2 consumption by target tissues and, accordingly, increases cardiac output and ventilation rate to match the increased O_2 consumption. Thyroid hormones cause increased heat production as a result of increased aerobic metabolism.

20–E. Hyperosmolarity is a specific stimulus for secretion of the other posterior pituitary hormone, ADH. Oxytocin is a powerful stimulant for contraction of the myoepithelial cells of the breast (ejecting milk) and of the uterine smooth muscle. Thus, the stimuli for its secretion are suckling, orgasm, and dilation of the cervix (as labor progresses).

21–E. GnRH is a peptide hormone that acts immediately upon the cells of the anterior pituitary to cause secretion of FSH and LH. 1,25-dihydroxycholecalciferol, progesterone, estradiol, and aldosterone are all steroid hormone derivatives of cholesterol and induce synthesis of new proteins.

22–B. Lack of insulin causes hyperglycemia (due to inability of cells to utilize glucose) and, as a result, glycosuria. There will be increased blood levels of fatty acids and their by-products, the ketoacids, which will produce metabolic acidosis, not metabolic alkalosis. The patient will have hyperkalemia caused by lack of insulin (decreased K^+ uptake into cells) and hyperventilation (respiratory compensation for metabolic acidosis).

23–D. During the second and third trimesters, the fetal adrenal gland synthesizes dehydroepiandrosterone, which is hydroxylated in the fetal liver and then transferred to the placenta where it is aromatized to estrogen. In the first trimester of pregnancy, the corpus luteum is the source of both estrogen and progesterone.

24–A. Decreased blood volume stimulates secretion of renin (because of decreased renal perfusion pressure) and initiates the renin-angiotensin-aldosterone cascade for stimulation of aldosterone secretion. Angiotensin-converting enzyme inhibitors block the cascade by decreasing production of angiotensin II. Hyperosmolarity stimulates ADH (not aldosterone) secretion. Hyperkalemia has a direct effect on the adrenal cortex to stimulate aldosterone secretion (which will then cause increased K^+ secretion by the kidney).

25–B. T_4 is deiodinated by iodinase at the level of the target tissue to produce T_3 (the more biologically active form). The alternative product of T_4 metabolism is reverse T_3, which is inactive.

26–D. In Graves' disease (hyperthyroidism), the thyroid is stimulated to produce and secrete vast quantities of thyroid hormones as a result of stimulation by thyroid-stimulating immunoglobulins (antibodies to the TSH receptors on the thyroid gland). Because of the high circulating levels of thyroid hormones, the anterior pituitary secretion of TSH will be turned off (negative feedback).

27–E. In order for I^- to be "organified" (incorporated into thyroid hormone), it must be oxidized to I_2, which is accomplished by a peroxidase enzyme in the thyroid follicular cell membrane. Propylthiouracil inhibits the peroxidase and, therefore, halts the synthesis of thyroid hormones.

28–A. Cholesterol conversion to pregnenolone is catalyzed by cholesterol desmolase. This step in the biosynthetic pathway for steroid hormones is stimulated by ACTH.

29–C. The conversion of 17-hydroxypregnenolone to dehydroepiandrosterone (as well as the conversion of 17-hydroxyprogesterone to androstenedione) is catalyzed by 17,20-lyase. If inhibited, processing to the androgens is stopped.

30–D. Aromatase is present in high concentrations in the ovarian granulosa cells. The adjacent theca cells produce mostly testosterone, which is transferred to the granulosa cells for aromatization to estradiol.

31–E. Target tissues for androgens contain 5α-reductase, which converts testosterone to dihydrotestosterone, the active form.

32–C. Somatostatin is secreted by the delta cells of the islets of Langerhans and acts locally (paracrine) to inhibit secretion by the alpha cells (glucagon) and by the beta cells (insulin).

33–A. The insulin receptor in target tissues is a tetramer. The two β subunits have tyrosine kinase activity and they autophosphorylate when stimulated by insulin.

34–D. Pancreatic lipase is secreted by the exocrine (not the endocrine) pancreas.

Comprehensive Examination

Directions: Each of the numbered items or incomplete statements in this section is followed by answers or by completions of the statement. Select the **one** lettered answer or completion that is **best** in each case.

Questions 1 and 2

After extensive testing, a patient is found to have a pheochromocytoma that secretes mainly epinephrine.

1. All of the following signs would be expected in this patient EXCEPT

(A) increased heart rate
(B) increased arterial blood pressure
(C) increased excretion rate of 3-methoxy-4-hydroxymandelic acid (VMA)
(D) flushing of the skin

2. Symptomatic treatment would be best achieved in this patient with

(A) phentolamine
(B) isoproterenol
(C) a combination of phentolamine and isoproterenol
(D) a combination of phentolamine and propranolol
(E) a combination of isoproterenol and phenylephrine

3. The principle of positive feedback is illustrated by the effect of

(A) P_{O_2} on breathing rate
(B) glucose on insulin secretion
(C) estrogen on FSH and LH secretion in midcycle
(D) blood $[Ca^{2+}]$ on parathyroid hormone secretion
(E) decreased blood pressure on sympathetic outflow to the heart and blood vessels

4. In the graph below, the response shown by the dotted line illustrates the effect of

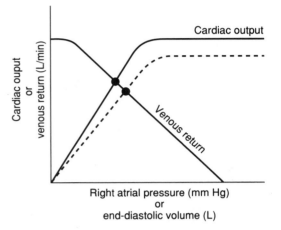

(A) administration of digitalis
(B) administration of a negative inotropic agent
(C) increased blood volume
(D) decreased blood volume
(E) decreased TPR

Questions 5 and 6

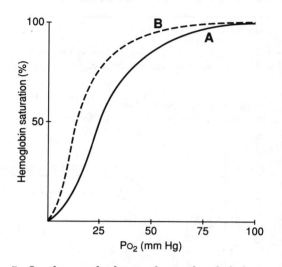

5. On the graph shown above, the shift from curve A to curve B could be caused by

(A) fetal hemoglobin (Hb F)
(B) carbon monoxide poisoning
(C) decreased pH
(D) increased temperature
(E) increased 2,3-DPG

6. The shift from curve A to curve B is associated with

(A) decreased P_{50}
(B) decreased affinity of hemoglobin for O_2
(C) decreased O_2-carrying capacity of hemoglobin
(D) increased ability to unload O_2 in the tissues

7. A negative C_{H_2O} (positive $T^c_{H_2O}$) would occur in a person

(A) who drinks 2 L of water in 30 minutes
(B) following overnight water restriction
(C) receiving lithium for treatment of depression who develops polyuria unresponsive to ADH administration
(D) with a urine flow rate of 5 ml/min, a urine osmolarity of 295 mOsm/L, and a serum osmolarity of 295 mOsm/L
(E) with a urine osmolarity of 90 mOsm/L and a serum osmolarity of 310 mOsm/L following a severe head injury

8. CO_2 generated in the tissues is carried in venous blood primarily as

(A) CO_2 in the plasma
(B) H_2CO_3 in the plasma
(C) HCO_3^- in the plasma
(D) CO_2 in the red blood cells
(E) carboxyhemoglobin in the red blood cells

9. In a 35-day menstrual cycle, ovulation occurs on day

(A) 12
(B) 14
(C) 17
(D) 21
(E) 28

10. Somatostatin inhibits the secretion of all of the following EXCEPT

(A) ADH
(B) gastrin
(C) glucagon
(D) insulin
(E) growth hormone

11. Which of the following is an example of a primary active transport process?

(A) Na^+–glucose transport in small intestinal epithelial cells
(B) Na^+–alanine transport in renal proximal tubular cells
(C) Insulin-dependent glucose transport in muscle cells
(D) H^+–K^+ transport in gastric parietal cells
(E) Ca^{2+}–Na^+ exchange in nerve cells

Questions 12 and 13

A patient with multiple myeloma is hospitalized after 2 days of polyuria, polydipsia, and increasing confusion. Laboratory tests show an elevated serum $[Ca^{2+}]$ of 15 mg/dl, and treatment is initiated to lower it. The patient's serum osmolarity is 310 mOsm/L.

12. The most likely reason for polyuria in this patient is

(A) increased circulating levels of ADH
(B) increased circulating levels of aldosterone
(C) inhibition of the action of ADH on the renal tubule
(D) stimulation of the action of ADH on the renal tubule
(E) psychogenic water drinking

13. Any of the following agents could be administered to reduce the patient's serum $[Ca^{2+}]$ EXCEPT

(A) thiazide diuretics
(B) loop diuretics
(C) calcitonin
(D) mithramycin
(E) etidronate disodium

14. Which of the following substances acts on its target cells via an IP_3–Ca^{2+} mechanism?

(A) Somatomedins acting on chondrocytes
(B) Oxytocin acting on myoepithelial cells of the breast
(C) ADH acting on the renal collecting duct
(D) ACTH acting on the adrenal cortex
(E) Thyroid hormone acting on muscle

15. A key difference in the mechanism of excitation–contraction coupling between the muscle of the pharynx and the muscle of the small intestine is that

(A) slow waves are present in the pharynx but not in the small intestine
(B) ATP is utilized for contraction in the pharynx but not in the small intestine
(C) there is an increase in intracellular $[Ca^{2+}]$ following excitation in the pharynx but not in the small intestine
(D) action potentials depolarize the muscle of the small intestine but not of the pharynx
(E) Ca^{2+} binds to troponin C in the pharynx but not in the small intestine to initiate contraction

16. A patient has an arterial pH of 7.25, an arterial P_{CO_2} of 30 mm Hg, and serum $[K^+]$ of 2.8 mEq/L. His blood pressure is 100/80 when supine and 80/50 when standing. What is the cause of this patient's abnormal blood values?

(A) Vomiting
(B) Diarrhea
(C) Treatment with a loop diuretic
(D) Treatment with a thiazide diuretic

17. Secretion of HCl by the gastric parietal cells is needed for

(A) activation of pancreatic lipases
(B) activation of salivary lipases
(C) activation of intrinsic factor
(D) activation of pepsinogen to pepsin
(E) formation of micelles

18. Which of the following would cause an increase in GFR?

(A) Constriction of the afferent arteriole
(B) Constriction of the efferent arteriole
(C) Constriction of the ureter
(D) Increased plasma protein concentration
(E) Infusion of inulin

19. Fat absorption occurs mainly in the

(A) stomach
(B) duodenum
(C) terminal ileum
(D) cecum
(E) sigmoid colon

20. Each of the following substances is converted to a more active form after its secretion EXCEPT

(A) testosterone
(B) T_4
(C) pepsinogen
(D) angiotensin I
(E) aldosterone

21. A 30-year-old woman has the anterior lobe of the pituitary gland surgically removed because of a tumor. Without hormone replacement therapy, which of the following would occur after the operation?

(A) Absence of menses
(B) Inability to concentrate the urine in response to water deprivation
(C) Failure to secrete catecholamines in response to stress
(D) Failure to secrete insulin in a glucose tolerance test
(E) Failure to secrete PTH in response to hypocalcemia

22. The graph below shows three relationships as a function of plasma [glucose]. At plasma [glucose] less than 200 mg/dl, curves X and Z are superimposed on each other because

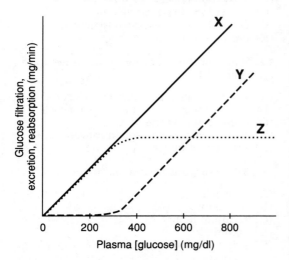

(A) reabsorption and excretion of glucose are equal
(B) all of the filtered glucose is reabsorbed
(C) glucose reabsorption is saturated
(D) the renal threshold for glucose has been exceeded
(E) Na^+–glucose cotransport has been inhibited

23. Which of the following responses occurs as a result of tapping on the patellar tendon?

(A) Stimulation of Ib afferent fibers in the muscle spindle
(B) Inhibition of Ia afferent fibers in the muscle spindle
(C) Relaxation of the quadriceps muscle
(D) Contraction of the quadriceps muscle
(E) Inhibition of α-motoneurons

Questions 24 and 25

A 5-year-old child has a severe sore throat, high fever, and cervical adenopathy.

24. It is suspected that the causative agent is *Streptococcus pyogenes*. Which of the following is involved in producing fever in the child?

(A) Increased production of interleukin-1 (IL-1)
(B) Decreased production of prostaglandins
(C) Decreased set-point temperature in the hypothalamus
(D) Decreased metabolic rate
(E) Vasodilation of blood vessels in the skin

25. Prior to the initiation of antibiotic therapy, the child is given aspirin to reduce his fever. The mechanism of fever reduction by aspirin is

(A) shivering
(B) stimulation of cyclo-oxygenase
(C) inhibition of prostaglandin synthesis
(D) shunting of blood from the surface of skin
(E) increasing the hypothalamic set-point temperature

26. A receptor potential in the pacinian corpuscle

(A) is "all-or-none"
(B) has a stereotypical size and shape
(C) is the action potential of this sensory receptor
(D) if hyperpolarizing, increases the likelihood of action potential occurrence
(E) if depolarizing, brings the membrane potential closer to threshold

27. In a skeletal muscle capillary, the capillary hydrostatic pressure is 32 mm Hg, the capillary π is 27 mm Hg, and the interstitial hydrostatic pressure is 2 mm Hg. Interstitial π is negligible. What is the driving force across the capillary wall and will it favor filtration or absorption?

(A) 3 mm Hg, favoring absorption
(B) 3 mm Hg, favoring filtration
(C) 7 mm Hg, favoring absorption
(D) 7 mm Hg, favoring filtration
(E) 9 mm Hg, favoring filtration

28. Each of the following conditions predisposes the formation of edema EXCEPT

(A) arteriolar constriction
(B) venous constriction
(C) standing
(D) nephrotic syndrome
(E) inflammation

29. A patient's ECG shows periodic QRS complexes that are not preceded by P waves and that have a bizarre shape. These QRS complexes originated in the

(A) SA node
(B) AV node
(C) His-Purkinje system
(D) ventricular muscle

30. Which of the following changes occurs during moderate exercise?

(A) Increased TPR
(B) Increased stroke volume
(C) Decreased pulse pressure
(D) Decreased venous return
(E) Decreased arterial P_{O_2}

Questions 31 and 32

A 17-year-old boy is brought to the emergency department after being injured in an automobile accident. He is given a transfusion of 3 units of blood to stabilize his blood pressure.

31. Prior to the transfusion, which of the following was true about his condition?

(A) His TPR was decreased
(B) His heart rate was decreased
(C) The firing rate of his carotid sinus nerves was increased
(D) There was increased sympathetic outflow to his heart and blood vessels

32. Each of the following is a consequence of the decrease in blood volume in this patient EXCEPT

(A) increased renal perfusion pressure
(B) increased renin secretion
(C) increased circulating levels of angiotensin II
(D) increased circulating levels of aldosterone
(E) increased renal Na^+ reabsorption

33. Each of the following lung capacities includes the tidal volume EXCEPT

(A) total lung capacity (TLC)
(B) vital capacity (VC)
(C) inspiratory capacity (IC)
(D) functional residual capacity (FRC)

34. Hypoxia causes vasodilation in which of the following vascular beds?

(A) Cerebral
(B) Coronary
(C) Muscle
(D) Pulmonary
(E) Skin

35. Hypoxemia will occur in each of the following situations EXCEPT

(A) a person having a severe asthmatic attack
(B) a person residing at high altitude
(C) a person with a reduced hemoglobin concentration
(D) a person with a left-to-right cardiac shunt
(E) a person with pulmonary fibrosis

36. Which of the following would be expected to increase after surgical removal of the duodenum?

(A) Gastric emptying time
(B) Secretion of CCK
(C) Secretion of secretin
(D) Contraction of the gallbladder
(E) Absorption of lipids

37. Each of the following is absorbed by Na^+- dependent cotransport (symport) EXCEPT

(A) glucose in duodenal cells
(B) fructose in duodenal cells
(C) dipeptides in duodenal cells
(D) vitamin B_1 in duodenal cells
(E) bile acids in ileal cells

Questions 38–41

Answer the following questions using the diagram of an action potential.

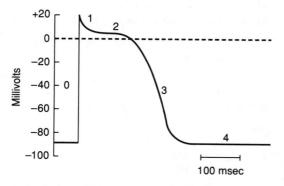

100 msec

38. The action potential depicted is from

(A) a skeletal muscle cell
(B) a smooth muscle cell
(C) a sinoatrial cell
(D) an atrial muscle cell
(E) a ventricular muscle cell

39. Phase 0 of the action potential is produced by an

(A) inward K^+ current
(B) inward Na^+ current
(C) inward Ca^{2+} current
(D) outward Na^+ current
(E) outward Ca^{2+} current

40. Phase 2, the plateau phase, of the action potential

(A) is the result of Ca^{2+} flux out of the cell
(B) increases in duration as heart rate increases
(C) corresponds to the effective refractory period
(D) is the result of approximately equal inward and outward currents
(E) is the portion of the action potential when another action potential can most easily be elicited

41. The action potential corresponds to which portion of an ECG?

(A) P wave
(B) PR interval
(C) QRS complex
(D) ST segment
(E) QT interval

42. Which of the following actions occurs when light strikes a photoreceptor cell of the retina?

(A) Transducin is inhibited
(B) The receptor cell depolarizes
(C) Cyclic GMP levels in the cell decrease
(D) *Trans*-rhodopsin is converted to *cis*-rhodopsin
(E) An excitatory neurotransmitter is always released

43. When compared with the base of the lung in a person who is standing, the apex of the lung has

(A) a higher ventilation rate
(B) a higher perfusion rate
(C) a higher V/Q ratio
(D) the same V/Q ratio
(E) a lower pulmonary capillary P_{O_2}

44. Which of the following substances has the lowest renal clearance in a person eating an "average" diet?

(A) Creatinine
(B) Glucose
(C) K^+
(D) Na^+
(E) PAH

45. Each of the following conditions causes hyperventilation EXCEPT

(A) strenuous exercise
(B) ascent to high altitude
(C) anemia
(D) diabetic ketoacidosis
(E) chronic obstructive pulmonary disease (COPD)

46. Which of the following substances would be expected to cause an increase in arterial blood pressure?

(A) Saralasin
(B) V_1 agonist
(C) ACh
(D) Spironolactone
(E) Phenoxybenzamine

47. Plasma renin activity is lower than normal in patients with

(A) hemorrhagic shock
(B) essential hypertension
(C) congestive heart failure
(D) hypertension caused by aortic constriction above the renal arteries

Directions: Each group of items in this section consists of lettered options followed by a set of numbered items. For each item, select the **one** lettered option that is most closely associated with it. Each lettered option may be selected once, more than once, or not at all.

Questions 48–53

Match the numbered laboratory and physical findings below with the correct patient description.

(A) Patient with chronic diabetic ketoacidosis
(B) Patient with chronic renal failure
(C) Patient with chronic emphysema and bronchitis
(D) Patient who hyperventilates on a commuter flight
(E) Patient taking a carbonic anhydrase inhibitor for glaucoma
(F) Patient with a pyloric obstruction who vomits for 5 days
(G) Healthy person

48. Arterial pH, 7.52; arterial P_{CO_2}, 26 mm Hg; tingling and numbness in the feet and hands

49. Arterial pH, 7.29; arterial $[HCO_3^-]$, 14 mEq/L; increased urinary excretion of NH_4^+; hyperventilation

50. Arterial $[HCO_3^-]$, 18 mEq/L; P_{CO_2}, 34 mm Hg; increased urinary HCO_3^- excretion

51. Arterial pH, 7.54; arterial $[HCO_3^-]$, 48 mEq/L; hypokalemia; hypoventilation

52. Arterial pH, 7.25; arterial P_{CO_2}, 30 mm Hg; decreased urinary excretion of NH_4^+

53. Arterial P_{CO_2}, 72 mm Hg; arterial $[HCO_3^-]$, 38 mEq/L; increased H^+ excretion

Questions 54–57

Match each numbered description below with the correct receptor.

(A) α Receptor
(B) β_1 Receptor
(C) β_2 Receptor
(D) Muscarinic receptor
(E) Nicotinic receptor

54. Albuterol is useful in the treatment of asthma because it acts as an agonist at this receptor

55. Activation of this receptor increases TPR

56. Atropine causes dry mouth by inhibiting this receptor

57. This receptor is present on the motor end plate of the neuromuscular junction

Questions 58–61

Match each numbered description below with the correct form of Na^+ transport.

(A) Na^+ diffusion via Na^+ channels
(B) Na^+–glucose cotransport (symport)
(C) Na^+–K^+–2 Cl^- cotransport (symport)
(D) Na^+–H^+ exchange (antiport)
(E) Na^+–K^+ ATPase

58. Inhibited by furosemide in the thick ascending limb

59. Results in the reabsorption of filtered HCO_3^- by the renal proximal tubule

60. An example of primary active transport

61. Involved in the upstroke of the action potential in nerves

Questions 62–73

Match each numbered description below with the correct hormone associated with it.

(A) ACTH
(B) ADH
(C) Aldosterone
(D) Dopamine
(E) 1,25-Dihydroxycholecalciferol
(F) Estradiol
(G) FSH
(H) GnRH
(I) HCG
(J) Insulin
(K) Oxytocin
(L) Prolactin
(M) PTH
(N) Somatostatin
(O) Testosterone

62. This hormone is converted to its active form in target tissues by the action of 5α-reductase

63. Deficiency of this hormone causes hypocalcemia and hyperphosphatemia

64. This hormone stimulates the conversion of cholesterol to pregnenolone

65. The receptor for this hormone has tyrosine kinase activity

66. Bromocriptine reduces galactorrhea by acting as an agonist for this substance

67. This hormone is derived from proopiomelanocortin (POMC)

68. This hormone acts on the anterior lobe of the pituitary to inhibit secretion of growth hormone

69. End-organ resistance to this hormone results in polyuria and an elevated serum osmolarity

70. This hormone is secreted in response to dilation of the cervix

71. This hormone stimulates conversion of testosterone to 17β-estradiol in ovarian granulosa cells

72. This hormone causes constriction of vascular smooth muscle through an IP_3 second messenger system

73. Levels of this hormone are high during the first trimester of pregnancy and subsequently decline during the second and third trimesters of pregnancy

Questions 74–78

Match each numbered phenomenon below with the correct lettered point on the ECG.

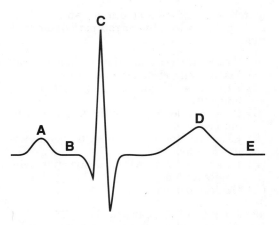

74. The atria are contracting

75. Both the atria and the ventricles are completely repolarized

76. The mitral valve closes

77. Aortic pressure is at its lowest value

78. Decreased duration of this portion of the ECG causes a decrease in pulse pressure

Questions 79–85

Match each numbered physiologic function below with the correct lettered site on the nephron.

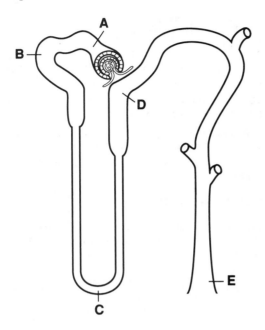

79. The amount of PAH in tubular fluid is the lowest

80. The creatinine concentration is highest in a person who is deprived of water

81. The tubular fluid $[HCO_3^-]$ is highest

82. The amount of K^+ in tubular fluid is lowest in a person on a diet very low in K^+

83. The composition of tubular fluid is closest to that of plasma

84. About one-third of the filtered water is remaining in the tubular fluid

85. The tubular fluid osmolarity is lower than the plasma osmolarity in a person deprived of water

Questions 86–88

Match each numbered description of a large systemic artery with the correct lettered parameter.

(A) Blood flow
(B) Resistance
(C) Pressure gradient
(D) Capacitance

86. If the artery is partially occluded by an embolism such that its radius becomes one-half the preocclusion value, this parameter will *increase* by a factor of 16

87. If the artery is partially occluded by an embolism such that its radius becomes one-half the preocclusion value, this parameter will *decrease* by a factor of 16

88. A decrease in this parameter will produce an increase in pulse pressure

Questions 89–93

Match each numbered description below with the correct phase of the left ventricular cardiac cycle.

(A) Atrial systole
(B) Isovolumetric ventricular contraction
(C) Rapid ventricular ejection
(D) Reduced ventricular ejection
(E) Isovolumetric ventricular relaxation
(F) Rapid ventricular filling
(G) Reduced ventricular filling

89. If heart rate increases, this phase is decreased

90. This phase is absent if there is no P wave on the ECG

91. During this phase, ventricular pressure is increasing and becomes slightly greater than aortic pressure

92. The second heart sound occurs during this phase

93. Phase during which ventricular pressure rises but ventricular volume is constant

Questions 94–97

Match each numbered patient description below with the correct diagnosis.

(A) Primary polydipsia
(B) Central diabetes insipidus
(C) Nephrogenic diabetes insipidus
(D) Water deprivation
(E) Syndrome of inappropriate ADH

94. A woman suffers a severe head injury in a skiing accident. Shortly thereafter, she becomes polydipsic and polyuric. Her urine osmolarity is 75 mOsm/L, and her serum osmolarity is 305 mOsm/L. Treatment with dDAVP causes an increase in her urine osmolarity to 450 mOsm/L.

95. A person has high circulating levels of ADH, a serum osmolarity of 260 mOsm/L, and a negative C_{H_2O}.

96. A patient who is receiving lithium treatment for bipolar depression becomes polyuric. His urine osmolarity is 90 mOsm/L; it remains at that level when he is given a nasal spray of dDAVP.

97. A person, who is thirsty, has a urine osmolarity of 950 mOsm/L and a serum osmolarity of 297 mOsm/L.

Questions 98–100

Match each numbered description below with the correct diuretic agent.

(A) Acetazolamide
(B) Chlorothiazide
(C) Furosemide
(D) Spironolactone
(E) Triamterene

98. Inhibits Na^+ reabsorption and K^+ secretion in the distal tubule by acting as an aldosterone antagonist

99. Administered for the treatment of acute mountain sickness and causes an increase in the pH of urine

100. Causes increased urinary excretion of Na^+ and K^+ and decreased urinary excretion of Ca^{2+}

Questions 101–104

Match each numbered description below with the correct GI secretory product.

(A) Saliva
(B) Gastric secretion
(C) Pancreatic secretion
(D) Bile

101. Has a component required for the intestinal absorption of vitamin B_{12}

102. Is hypotonic, has a high $[HCO_3^-]$, and its production is inhibited by vagotomy

103. Stimulated by secretin and contains the enzymes necessary for fat digestion

104. Inhibited when the pH of stomach contents is 1.0

Questions 105–107

Match each numbered description below with the correct hormone that has an action on the kidney.

(A) ADH
(B) Aldosterone
(C) ANF (atriopeptin)
(D) 1,25-Dihydroxycholecalciferol
(E) PTH

105. Secretion is stimulated by ECF volume expansion

106. Acts on the kidney to increase urinary phosphate excretion

107. Causes contraction of vascular smooth muscle

Questions 108–111

Match each numbered description below with the correct enzyme in the steroid hormone biosynthetic pathway.

(A) Aldosterone synthase
(B) Aromatase
(C) Cholesterol desmolase
(D) 17,20-Lyase
(E) 5α-Reductase

108. Stimulated by angiotensin II

109. Inhibition blocks production of all androgenic compounds, but does not block production of glucocorticoids or mineralocorticoids

110. Required for the development of female secondary sex characteristics but not for male secondary sex characteristics

111. Inhibition reduces the size of the prostate

Questions 112–115

Match each numbered description below with the correct step in the biosynthetic pathway for thyroid hormones.

(A) I^- pump
(B) $I^- \rightarrow I_2$
(C) I_2 + tyrosine
(D) DIT + DIT
(E) $T_4 \rightarrow T_3$

112. Inhibited by propylthiouracil

113. Produces T_4

114. Inhibited by thiocyanate

115. Catalyzed by 5'-iodinase

Answers and Explanations

1–D. Increased circulating levels of epinephrine from the adrenal medullary tumor stimulate both α-adrenergic and β-adrenergic receptors. Thus, heart rate and contractility are increased, resulting in increased cardiac output. TPR is increased because of arteriolar vasoconstriction; flushing would occur if arterioles in the skin were dilated, not constricted. Together, the increases in cardiac output and TPR produce an increase in arterial blood pressure. VMA is a metabolite of both norepinephrine and epinephrine; increases in VMA excretion occur in pheochromocytomas.

2–D. Treatment must be directed at blocking both the α-stimulatory and β-stimulatory effects of epinephrine. Phentolamine is an α-blocking agent and propranolol is a β-blocking agent.

3–C. The effect of estrogen on secretion of FSH and LH by the anterior lobe of the pituitary gland at midcycle is one of very few examples of positive feedback in physiologic systems—increasing estrogen levels at midcycle causes *increased* secretion of FSH and LH. The other options listed illustrate negative feedback. Decreased arterial P_{O_2} causes an increase in breathing rate (via peripheral chemoreceptors). Increased blood glucose stimulates insulin secretion. Decreased blood $[Ca^{2+}]$ causes an increase in parathyroid hormone secretion. Decreased blood pressure causes decreased firing rate of carotid sinus nerves (via the baroreceptors) and ultimately increases sympathetic outflow to the heart and blood vessels to increase blood pressure back to normal.

4–B. A downward shift of the cardiac output curve is consistent with decreased myocardial contractility (negative inotropism); for any right atrial pressure (sarcomere length), the force of contraction is decreased. Digitalis, a positive inotropic agent, would produce an upward shift of the cardiac output curve. Changes in blood volume alter the venous return curve, rather than the cardiac output curve. Changes in TPR alter both cardiac output and venous return.

5–A. Because fetal hemoglobin (Hb F) has a greater affinity for O_2 than does adult hemoglobin, the O_2–hemoglobin dissociation curve would *shift to the left*. Carbon monoxide poisoning would cause a shift to the left, but it would also cause a decrease in total O_2-carrying capacity (decreased percent saturation). Decreased pH, increased temperature, and increased 2,3-DPG would all shift the curve to the right.

6–A. A shift to the left of the O_2–hemoglobin dissociation curve represents an increased affinity of hemoglobin for O_2. Accordingly, at any given level of P_{O_2}, the percent saturation is increased, the P_{50} is decreased (read the P_{O_2} at 50% saturation), and the ability to unload O_2 to the tissues is impaired (higher affinity of hemoglobin for O_2). The O_2-carrying capacity is determined by hemoglobin concentration and is unaffected by the shift from curve A to curve B.

7–B. A person with a negative C_{H_2O} would, by definition, be producing urine that is hyperosmotic to blood ($C_{H_2O} = V - C_{osm}$). Following overnight water restriction, there is a rise in serum osmolarity which, via hypothalamic osmoreceptors, stimulates release of ADH from the posterior lobe of the pituitary. This ADH circulates to the collecting ducts of the kidney and causes reabsorption of water, resulting in the production of hyperosmotic urine. Drinking large amounts of water would inhibit secretion of ADH and cause excretion of dilute urine and a positive C_{H_2O}. Lithium would cause nephrogenic diabetes insipidus by blocking the response of ADH on the collecting duct cells, resulting in dilute urine and positive C_{H_2O}.

8–C. CO_2 generated in the tissues enters venous blood and, in the red blood cells, combines with H_2O in the presence of carbonic anhydrase to form H_2CO_3. H_2CO_3 dissociates into H^+ and HCO_3^-. The H^+ remains in the red blood cells buffered by deoxyhemoglobin, and the HCO_3^- moves into plasma in exchange for Cl^-. Thus, CO_2 is carried in venous blood to the lungs as HCO_3^-. In the lungs the reactions occur in reverse: CO_2 is regenerated and expired.

9–D. Menses occurs reliably 14 days after ovulation. Therefore, in a 35-day menstrual cycle, ovulation will occur on day 21. Note that ovulation occurs at the midpoint of the menstrual cycle only if the cycle length is 28 days.

10–A. The actions of somatostatin are diverse. It is secreted by the hypothalamus to inhibit the secretion of growth hormone by the anterior lobe of the pituitary. It is secreted by cells of the GI tract to inhibit the secretion of the GI hormones. It is also secreted by the delta cells of the endocrine pancreas and, via paracrine mechanisms, inhibits the secretion of insulin and glucagon by the beta cells and alpha cells, respectively.

11–D. H^+-K^+ transport in gastric parietal cells occurs via the H^+-K^+ ATPase in the luminal membrane, a primary active transport process energized directly by ATP. Na^+–glucose and Na^+–alanine transport are examples of cotransport (symport) that are secondary active and do not utilize ATP directly. Glucose uptake into muscle cells occurs via facilitated diffusion. Na^+-Ca^{2+} exchange is an example of countertransport (antiport) and is secondary active.

12–C. The most likely explanation for this patient's polyuria is his hypercalcemia. With severe hypercalcemia, Ca^{2+} accumulates in the inner medulla and papilla of the kidney and inhibits ADH-stimulated adenylate cyclase, blocking the effect of ADH on water permeability. As a result of the ineffectiveness of ADH, the urine cannot be concentrated and the patient excretes large volumes of dilute urine. His polydipsia is secondary to his polyuria, caused by the increased serum osmolarity. Psychogenic water drinking would also cause polyuria, but the serum osmolarity would be lower than normal, not higher than normal.

13–A. Thiazide diuretics would be contraindicated in a patient with severe hypercalcemia because these drugs cause increased Ca^{2+} reabsorption in the renal distal tubule. Loop diuretics, on the other hand, inhibit Ca^{2+} and Na^+ reabsorption and produce calciuresis; when given with fluid replacement, loop diuretics can effectively and rapidly lower the serum $[Ca^{2+}]$. Calcitonin, mithramycin, and etidronate disodium inhibit bone resorption and, as a result, decrease serum $[Ca^{2+}]$.

14–B. Oxytocin causes contraction of the myoepithelial cells of the breast by an IP_3–Ca^{2+} mechanism. Somatomedins (IGF), like insulin, act on target cells by activating tyrosine kinase. ADH acts on the V_2 receptors of the renal collecting duct by a cyclic AMP mechanism (although in vascular smooth muscle it acts on V_1 receptors by an IP_3 mechanism). ACTH also acts via a cyclic AMP mechanism. Thyroid hormone induces synthesis of new protein (e.g., Na^+-K^+ ATPase) by a steroid hormone mechanism.

15–E. Identify that the pharynx is skeletal muscle and the small intestine is unitary smooth muscle. Both types of muscle are excited to contract by action potentials. Slow waves are present in smooth muscle but not in skeletal muscle. Both smooth and skeletal muscle require an increase in intracellular $[Ca^{2+}]$ as the important linkage between excitation (the action potential) and contraction, and both consume ATP. The key difference between smooth and skeletal muscle is the mechanism by which Ca^{2+} initiates contraction—in smooth muscle Ca^{2+} binds to calmodulin, and in skeletal muscle Ca^{2+} binds to troponin C.

16–B. The arterial blood values and physical findings are consistent with metabolic acidosis, hypokalemia, and orthostatic hypotension. Diarrhea is associated with loss of HCO_3^- and K^+ from the GI tract, consistent with the laboratory values. The hypotension is consistent with ECF volume contraction. Vomiting would cause metabolic alkalosis and hypokalemia. Treatment with loop or thiazide diuretics *could* cause volume contraction and hypokalemia, but *would* cause metabolic alkalosis rather than metabolic acidosis.

17–D. Pepsinogen is secreted by the gastric chief cells and is activated to pepsin by the low pH of the stomach (achieved by secretion of HCl by the gastric parietal cells). Lipases are *inactivated* by acidity.

18–B. GFR is determined by the balance of Starling forces across the glomerular capillary wall. Constriction of the efferent arteriole increases the glomerular capillary hydrostatic pressure (since blood is restricted in leaving the glomerular capillary), thus favoring filtration. Constriction of the afferent arteriole would have the opposite effect and lower the glomerular capillary hydrostatic pressure. Constriction of the ureter would raise the hydrostatic pressure in the tubule and, therefore, oppose filtration. Increased plasma protein concentration would raise the glomerular capillary oncotic pressure and oppose filtration. Infusion of inulin is used to *measure* GFR and does not affect the Starling forces.

19–B. First, fat absorption requires the breakdown of dietary lipids to fatty acids, monoglycerides, and cholesterol in the duodenum by pancreatic lipases. Second, fat absorption requires the presence of bile acids, which are secreted into the small intestine by the gallbladder. These bile acids form micelles around the products of lipid digestion and deliver them to the absorbing surface of the small intestinal cells. Because the bile acids are recirculated to the liver from the ileum, fat absorption must be complete before the chyme reaches the terminal ileum.

20–E. Both testosterone and T_4 are converted to more active forms in their target tissues (to dihydrotestosterone and T_3, respectively). Pepsinogen is activated to pepsin, the active form of this proteolytic enzyme. Angiotensin I is converted to its active form, angiotensin II, by the action of angiotensin-converting enzyme. Aldosterone is unchanged after its secretion by the zona glomerulosa of the adrenal cortex.

21–A. Normal menstrual cycles depend upon the secretion of FSH and LH from the anterior pituitary. Concentration of urine in response to water deprivation depends upon secretion of ADH by the posterior pituitary. Catecholamines are secreted by the adrenal medulla in response to stress, but releasing hormones from the anterior pituitary are not involved. Anterior pituitary hormones are not involved in the direct effect of glucose on the beta cells of the pancreas nor in the direct effect of Ca^{2+} on the chief cells of the parathyroid gland.

22–B. Curves X, Y, and Z can be identified as glucose filtration, glucose excretion, and glucose reabsorption, respectively. Below a plasma [glucose] of 200 mg/dl, the carriers for glucose reabsorption are unsaturated, so all of the filtered glucose can be reabsorbed and none will be excreted in the urine.

23–D. When the patellar tendon is stretched, the quadriceps muscle stretches. This movement activates Ia afferent fibers of the muscle spindles, which are arranged in parallel formation in the muscle. These Ia afferent fibers form synapses on α-motoneurons in the spinal cord. In turn, the pool of α-motoneurons is activated and causes reflex contraction of the quadriceps muscle to return it to its resting length.

24–A. *Streptococcus pyogenes* causes increased production of IL-1 in macrophages. IL-1 acts on the anterior hypothalamus to increase production of prostaglandins, which increase the hypothalamic set-point temperature. The hypothalamus then "reads" the core temperature as being lower than the new set-point temperature, and activates various heat-generating mechanisms that raise body temperature (fever). These mechanisms include shivering and vasoconstriction of blood vessels in the skin.

25–C. By inhibiting cyclo-oxygenase, aspirin inhibits the production of prostaglandins and brings the hypothalamic set point back down to its original value. After aspirin treatment, the hypothalamus "reads" the body temperature as being higher than the set point, and activates heat-loss mechanisms, including sweating and vasodilation of skin blood vessels (which shunts blood toward the surface skin). When heat is lost from the body by these mechanisms, body temperature is reduced.

26–E. Receptor potentials in sensory receptors (such as pacinian corpuscle) are not action potentials and do not have the stereotypical size and shape, or the all-or-none feature of the action potential. Instead, they are graded potentials that vary in size depending on stimulus intensity. A hyperpolarizing receptor potential would take the membrane potential away from threshold and decrease the likelihood of action potential occurrence. A depolarizing receptor potential would bring the membrane potential toward threshold and increase the likelihood of action potential occurrence.

27–B. The driving force is calculated from the Starling forces across the capillary wall. The net pressure = $(P_c - P_i) - (\pi_c - \pi_i)$. Therefore, net pressure = (32 mm Hg − 2 mm Hg) − (27 mm Hg) = 3 mm Hg. Since the sign of the net pressure is positive, filtration is favored.

28–A. Constriction of arterioles causes decreased capillary hydrostatic pressure and, as a result, decreased net pressure (Starling forces) across the capillary wall. Filtration will be reduced, as will the tendency for edema to form. Venous constriction and standing cause increased capillary hydrostatic pressure, and tend to cause filtration and edema formation. Nephrotic syndrome results in the excretion of plasma proteins in the urine and a decrease in the oncotic pressure of capillary blood, which also predisposes filtration and edema formation. Inflammation causes local edema by causing dilation of arterioles.

29–C. Since there are no P waves associated with the bizarre QRS complex, activation could not have begun in the SA node. Had the beat originated in the AV node, the QRS complex would have had a "normal" shape because the ventricles would activate in their normal sequence. The beat must have originated in the His-Purkinje system, and the bizarre shape of the QRS complex reflects an improper activation sequence of the ventricles. Ventricular muscle does not have pacemaker properties.

30–B. During moderate exercise, there is increased sympathetic outflow to the heart and blood vessels. The sympathetic effects on the heart cause increased heart rate and contractility, and the increased contractility results in increased stroke volume. Pulse pressure increases as a consequence of the increased stroke volume. Venous return also increases because of muscular activity; this increased venous return further contributes to increased stroke volume by the Frank-Starling mechanism. TPR might be expected to increase because of sympathetic stimulation of the blood vessels. However, the buildup of local metabolites in the exercising muscle causes local vasodilation, which overrides the sympathetic vasoconstrictor effect (thus decreasing TPR). *Arterial P_{O_2} does not decrease* during moderate exercise although O_2 consumption increases.

31–D. The blood loss that occurred in the accident caused an immediate loss of volume from the vascular compartment and an associated decrease in arterial blood pressure. The fall in arterial pressure was detected by the baroreceptors in the carotid sinus and resulted in decreased firing rate of the carotid sinus nerves. As a result of the baroreceptor response, sympathetic outflow to the heart and blood vessels increased, and parasympathetic outflow to the heart decreased, producing together an increased heart rate, increased contractility, and increased TPR (all trying to restore the arterial blood pressure).

32–A. The decreased blood volume causes *decreased* renal perfusion pressure, which initiates a cascade of events, including increased renin secretion, increased circulating angiotensin II, increased aldosterone secretion, and increased Na^+ reabsorption by the renal tubule.

33–D. The FRC is the volume remaining in the lungs after expiration of a normal tidal volume.

34–B. Both pulmonary and coronary circulations are regulated by the P_{O_2}. However, the critical difference is that hypoxia causes vasodilation in the coronary circulation and causes vasoconstriction in the pulmonary circulation. Cerebral and muscle circulations are regulated primarily by local metabolites, and skin circulation is regulated primarily by sympathetic innervation (for temperature regulation).

35–D. A person with a left-to-right cardiac shunt will have an elevated P_{O_2} on the right side of the heart because arterial blood is mixed with venous blood. During an asthmatic attack, a person may have a reduced P_{O_2} because of increased resistance to airflow. At high altitude, arterial P_{O_2} is reduced because the inspired air has a reduced P_{O_2}. Persons with reduced hemoglobin concentrations have decreased O_2-carrying capacity. In pulmonary fibrosis, diffusion of O_2 across the alveolar membrane is decreased.

36–A. Removal of the duodenum would remove the source of the GI hormones, CCK and secretin. Because CCK stimulates contraction of the gallbladder (and, therefore, ejection of bile acids into the intestine), lipid absorption would be impaired. CCK also inhibits gastric emptying, so removing the source of CCK should increase gastric emptying time.

37–B. The monosaccharides (glucose, galactose, and fructose) are the absorbable forms of carbohydrates. Only glucose and galactose are absorbed by Na^+–dependent cotransport; fructose is absorbed by facilitated diffusion. Dipeptides and water-soluble vitamins are absorbed by Na^+–dependent cotransport in the duodenum, and bile acids are absorbed by Na^+–dependent cotransport in the ileum (which recycles them to the liver).

38–E. Action potentials in skeletal cells are of much shorter duration than that depicted in the figure (only a few msec). Smooth muscle action potentials would be superimposed upon fluctuating baseline potentials (slow waves). Sinoatrial cells of the heart would have spontaneous depolarization (pacemaker activity) rather than a stable resting potential. Atrial muscle cells of the heart have a much shorter plateau phase and a much shorter overall duration. The action potential depicted is characteristic of ventricular muscle with a stable resting membrane potential and a plateau phase of almost 300 msec.

39–B. Depolarization, as in phase 0, is caused by an inward current (defined as the movement of positive charge into the cell). The inward current during phase 0 of the ventricular muscle action potential is caused by opening of Na^+ channels in the ventricular muscle cell membrane, movement of Na^+ into the cell, and depolarization of the membrane potential toward the Na^+ equilibrium potential (approximately $+65$ mV). Note that in sinoatrial cells phase 0 is caused by an inward Ca^{2+} current.

40–D. Because the plateau phase is a period of stable membrane potential, by definition the inward and outward currents are equal and balance each other. Phase 2 is the result of opening of Ca^{2+} channels and *inward Ca^{2+} flux*. In this phase, the cells are refractory to initiation of another action potential; however, phase 2 corresponds to the absolute refractory period, rather than the effective refractory period (which is longer than the plateau). As heart rate increases, the duration of the ventricular action potential decreases, primarily by decreasing the duration of phase 2.

41–E. The action potential depicted represents both depolarization and repolarization of a ventricular muscle cell. Therefore, on an ECG, it corresponds to the period of depolarization (beginning with the Q wave) through repolarization (completion of the T wave). That period is defined as the QT interval.

42–C. Light striking a photoreceptor cell causes conversion of *cis*-rhodopsin to *trans*-rhodopsin; activation of a G protein called transducin; activation of phosphodiesterase, which catalyzes the conversion of cyclic GMP to 5′-GMP so that *cyclic GMP levels decrease*; closure of Na^+ channels by the decreased cyclic GMP levels; hyperpolarization of the cell; and release of excitatory *or* inhibitory neurotransmitter.

43–C. In a person who is standing, both ventilation and perfusion are higher at the base of the lung than at the apex of the lung. However, since the regional differences for perfusion are greater than the regional differences for ventilation, the V/Q ratio is higher at the apex than at the base. The pulmonary capillary P_{O_2} is higher at the apex than at the base because the higher V/Q ratio makes gas exchange more efficient.

44–B. Glucose has the lowest renal clearance of the substances listed, because at normal blood concentrations it is filtered and completely reabsorbed. Na^+ is also extensively reabsorbed and only a small fraction of the filtered Na^+ is excreted. K^+ is reabsorbed but also secreted. Creatinine, once filtered, is not reabsorbed at all. PAH is filtered and secreted and, therefore, it has the highest renal clearance of the substances listed.

45–E. Strenuous exercise increases the ventilation rate to provide additional oxygen to the exercising muscle. Ascent to high altitude and anemia result in hypoxemia, which subsequently causes hyperventilation by stimulating peripheral chemoreceptors. The respiratory compensation for diabetic ketoacidosis is hyperventilation. COPD causes hypoventilation.

46–B. Because saralasin is an angiotensin-converting enzyme inhibitor, it blocks the production of the vasoconstrictor substance, angiotensin II. A V_1 agonist simulates the vasoconstrictor effects of ADH. Spironolactone, an aldosterone antagonist, blocks the effects of aldosterone to increase distal tubule Na^+ reabsorption, and consequently reduces ECF volume and blood pressure. Phenoxybenzamine, an α-blocking agent, inhibits the vasoconstrictor effect of α-adrenergic stimulation. ACh, via production of EDRF, causes vasodilation of vascular smooth muscle and reduces blood pressure.

47–B. Patients with essential hypertension have decreased renin secretion as a result of increased renal perfusion pressure. Patients with congestive heart failure and hemorrhagic shock have increased renin secretion because of reduced intravascular volume, which results in decreased renal perfusion pressure. Patients with aortic constriction above the renal arteries are hypertensive because decreased renal perfusion pressure causes increased renin secretion, followed by increased secretion of angiotensin II and aldosterone.

48–D. The blood values are consistent with acute respiratory alkalosis from hysterical hyperventilation. The tingling and numbness are symptoms of a reduction in serum ionized $[Ca^{2+}]$ that occurs as a result of alkalosis. Because of the reduction in $[H^+]$, fewer H^+ ions will bind to negatively charged sites on plasma proteins, and more Ca^{2+} will be bound (decreasing the free ionized $[Ca^{2+}]$).

49–A. The blood values are consistent with metabolic acidosis, such as would occur in diabetic ketoacidosis. Hyperventilation is the respiratory compensation for metabolic acidosis. Increased urinary excretion of NH_4^+ reflects the adaptive increase in NH_3 synthesis that occurs in chronic acidosis. (Note that patients with metabolic acidosis from chronic renal failure would have a reduced NH_4^+ excretion [because of diseased renal tissue], which *causes* metabolic acidosis.)

50–E. The blood values are consistent with metabolic acidosis (calculate pH = 7.34). Treatment with a carbonic anhydrase inhibitor is suggested because it would increase HCO_3^- excretion.

51–F. The blood values and history of vomiting are consistent with metabolic alkalosis. Hypoventilation is the respiratory compensation for metabolic alkalosis. Hypokalemia results from loss of gastric K^+ and from hyperaldosteronism secondary to volume contraction.

52–B. The blood values are consistent with metabolic acidosis with respiratory compensation. Because the urinary excretion of NH_4^+ is decreased, chronic renal failure is a likely cause.

53–C. The blood values are consistent with respiratory acidosis with renal compensation. The renal compensation involves increased reabsorption of HCO_3^- (associated with increased H^+ secretion), which raises the serum $[HCO_3^-]$.

54–C. Albuterol is an adrenergic β_2 agonist. The bronchioles have β_2 receptors, which, when activated, produce bronchodilation.

55–A. When adrenergic α receptors on the vascular smooth muscle are activated, they cause vasoconstriction and increased TPR.

56–D. Atropine blocks cholinergic muscarinic receptors. Because saliva production is increased by stimulation of the parasympathetic nervous system, atropine treatment will reduce saliva production and cause dry mouth.

57–E. Nicotinic receptors for ACh are present on the motor end plate of the neuromuscular junction, in autonomic ganglia, and in the adrenal gland.

58–C. Na^+–K^+–$2\,Cl^-$ cotransport is the mechanism in the luminal membrane of thick ascending limb cells that is inhibited by loop diuretics such as furosemide. Other loop diuretics that inhibit this transporter are bumetanide and ethacrynic acid.

59–D. Na^+–H^+ exchange in the luminal membrane of the early proximal tubular cells transports H^+ into the lumen, which subsequently combines with filtered HCO_3^- to form H_2CO_3. H_2CO_3 dissociates into CO_2 and H_2O, which move into the proximal tubular cell. Within the cell, CO_2 and H_2O recombine into H_2CO_3, which dissociates into H^+ and HCO_3^-. The HCO_3^- (originally *filtered* HCO_3^-) is reabsorbed into blood, and the H^+ is used again by the Na^+–H^+ exchange transport process to aid in the reabsorption of more filtered HCO_3^-.

60–E. Primary active transport requires a direct supply of metabolic energy in the form of ATP; for example, the Na^+–K^+ pump (or Na^+–K^+ ATPase); others examples are the H^+–K^+ ATPase in gastric parietal cells and Ca^{2+}–ATPase in sarcoplasmic reticulum. Na^+ channels allow the transport of Na^+ by simple diffusion down an electrochemical gradient, requiring no metabolic energy. Cotransport (symport) and antiport processes are secondary active and use energy derived from the Na^+ gradient that exists across cell membranes.

61–A. During the upstroke of the action potential, Na^+ channels are opened by depolarization. When the Na^+ channels open, Na^+ rushes into nerve cells by diffusion down its electrochemical gradient and drives the membrane potential toward the Na^+ equilibrium potential, thus further depolarizing the nerve cell.

62–O. Testosterone is converted to its active form, dihydrotestosterone, in its target tissues by the action of 5α-reductase. Deficiency of this enzyme can prevent development of male secondary sex characteristics and prevent development of the accessory sex organs (prostate, epididymis, seminal vesicles).

63–M. Lack of PTH (hypoparathyroidism) causes decreased bone resorption, decreased renal Ca^{2+} reabsorption, increased renal phosphate reabsorption, and decreased production of the active form of vitamin D (1,25-dihydroxycholecalciferol). Together, these effects lower the serum $[Ca^{2+}]$ and increase the serum [phosphate].

64–A. In the adrenal cortex, ACTH stimulates cholesterol desmolase, which catalyzes conversion of cholesterol to pregnenolone, thereby initiating the entire steroid hormone biosynthetic pathway.

65–J. Hormone receptors with tyrosine kinase activities include those for insulin and for insulin-like growth factors (IGF). The β subunits of the insulin receptor have tyrosine kinase activity and, when activated by insulin, the receptors autophosphorylate. The phosphorylated receptors then phosphorylate intracellular proteins; this process ultimately results in the physiologic actions of insulin.

66–D. Prolactin secretion by the anterior pituitary is tonically inhibited by dopamine secreted by the hypothalamus. If this inhibition is disrupted (e.g., by interruption of the hypothalamic–pituitary tract), then prolactin secretion will be increased, causing galactorrhea. The dopamine agonist bromocriptine simulates the tonic inhibition by dopamine and inhibits prolactin secretion.

67–A. In the anterior pituitary, proopiomelanocortin is the precursor molecule to ACTH, β-lipotropin, and β-endorphin.

68–N. Somatostatin is secreted by the hypothalamus and inhibits the secretion of growth hormone by the anterior pituitary. Notably, much of the feedback inhibition of growth hormone secretion occurs by stimulating the secretion of somatostatin (an inhibitory hormone). Both growth hormone itself and somatomedins (IGF; produced in the target tissues for growth hormone) stimulate the secretion of somatostatin by the hypothalamus.

69–B. End-organ resistance to ADH is called nephrogenic diabetes insipidus. It may be caused by lithium intoxication (which inhibits the G_s protein in collecting duct cells) or by hypercalcemia (which inhibits adenylate cyclase). The result is inability to concentrate the urine, polyuria, dilute urine, and increased serum osmolarity (resulting from loss of free water in the urine).

70–K. Oxytocin is secreted by the posterior pituitary in response to dilation of the cervix and suckling.

71–G. Testosterone is synthesized from cholesterol in the ovarian theca cells and diffuses to the ovarian granulosa cells where it is converted to estradiol by the action of aromatase. FSH stimulates the aromatase enzyme and increases the production of estradiol.

72–B. ADH causes constriction of vascular smooth muscle by activating a V_1 receptor that utilizes the IP_3 and Ca^{2+} second messenger system. When there is hemorrhage or ECF volume contraction, ADH secretion by the posterior pituitary is stimulated via volume receptors. The resulting increase in ADH levels causes increased water reabsorption by the collecting ducts (V_2 receptors) and vasoconstriction (V_1 receptors) to help restore the blood pressure.

73–I. During the first trimester of pregnancy, the placenta produces HCG, which then stimulates estrogen and progesterone production by the corpus luteum. Peak levels of HCG occur at about the ninth gestational week and then decline. At the time of the decline in HCG, the placenta assumes the role of steroidogenesis for the remainder of the pregnancy.

74-B. The atria contract following their electrical activation (P wave). Atrial contraction contributes, in part, to ventricular filling and so must be complete before the ventricles contract.

75–E. The atria depolarize during the P wave and then repolarize. The ventricles depolarize during the QRS complex and then repolarize during the T wave. Thus, both atria and ventricles are fully repolarized at the completion of the T wave.

76–C. The mitral valve closes at the peak of ventricular activation (R wave) so that when the left ventricle contracts it can eject blood into the aorta, not back into the left atrium. Prior to ventricular contraction, the mitral valve remains open so that the ventricles can fill.

77–C. Aortic pressure is lowest just prior to contraction of the ventricles. At this point, the greatest amount of blood is in the heart rather than in the blood vessels.

78–B. A decrease in the duration of the PR interval results in decreased time available for filling of the ventricles from the atria. As a result, there is decreased stroke volume and decreased pulse pressure. (Recall that pulse pressure is determined by the volume of blood ejected into the arteries—that is, stroke volume.)

79–A. PAH is filtered across the glomerular capillaries and then secreted by the cells of the late proximal tubule. The sum of filtration plus secretion of PAH equals its excretion rate. Therefore, the smallest amount of PAH present in tubular fluid will be in the glomerular filtrate prior to the site of secretion.

80–E. Creatinine is a glomerular marker that serves a function similar to inulin. The creatinine concentration in tubular fluid is an indicator of water reabsorption along the nephron—the creatinine concentration rises as water is reabsorbed. In a person deprived of water (antidiuresis), water is reabsorbed throughout the nephron, including the collecting ducts, and the creatinine concentration is highest in the final urine.

81–A. HCO_3^- is filtered and then extensively reabsorbed in the early proximal tubule. Because this reabsorption exceeds that for H_2O reabsorption, the $[HCO_3^-]$ of proximal tubular fluid falls. Therefore, the highest concentration of $[HCO_3^-]$ will be found in the glomerular filtrate.

82–E. K^+ is filtered and then reabsorbed in the proximal tubule and loop of Henle. In a person on a diet very low in K^+, the distal tubule continues to reabsorb K^+ so that the amount of K^+ present in tubular fluid will be lowest in the final urine. (Note that if the person were on a diet high in K^+, then K^+ would be secreted, not reabsorbed, in the distal tubule.)

83–A. Tubular fluid is most like plasma in the glomerular filtrate; there, its composition is virtually identical to plasma, except that it does not contain plasma proteins (which are restricted from passage across the glomerular capillary because of their molecular size). Once the tubular fluid leaves Bowman's space, it is modified by the cells lining the tubule.

84–B. The proximal tubule reabsorbs about two-thirds of the glomerular filtrate isosmotically; thus, one-third of the glomerular filtrate remains at the end of the proximal tubule.

85–D. Under conditions of either water deprivation (antidiuresis) or water loading, the thick ascending limb of the loop of Henle performs its basic function of reabsorbing salt without water (due to the water impermeability of this segment). Thus, fluid leaving the loop of Henle is dilute with respect to plasma, even if the final urine is more concentrated than plasma.

86–B. A decrease in radius causes an increase in resistance, as described by the Poiseuille relationship (resistance is inversely proportional to r^4). Thus, if radius decreases 2-fold, the resistance will increase by $(2)^4$ or 16-fold.

87–A. Blood flow through the artery is proportional to the pressure difference and inversely proportional to the resistance ($Q = \Delta P/R$). Since resistance increased 16-fold when the radius decreased 2-fold, then blood flow must decrease 16-fold.

88–D. A decrease in capacitance of the artery means that for a given volume of blood in the artery, the pressure will be increased. Thus, for a given stroke volume ejected into the artery, the systolic pressure will be higher and the pulse pressure will also be higher.

89–G. When heart rate increases, there is decreased time between ventricular contractions for refilling of the ventricles with blood. Since the majority of ventricular filling occurs during the "reduced" phase, this phase will be most compromised by an increase in heart rate.

90–A. The P wave represents electrical activation (depolarization) of the atria. Atrial contraction is always preceded by electrical activation.

91–C. During the phase of rapid ventricular ejection, ventricular pressure is rising to its maximum value because of the contraction of the ventricle. The aortic valve opens when ventricular pressure just exceeds aortic pressure, allowing blood to be ejected into the aorta.

92–E. Closure of the aortic and pulmonic valves creates the second heart sound. The closure of these valves corresponds to the end of ventricular ejection and the beginning of ventricular relaxation.

93–B. Because the ventricles contract during this phase, the ventricular pressure rises. However, since all valves are closed, the contraction is isovolumetric. No blood will be ejected into the aorta until ventricular pressure rises enough to open the aortic valve.

94–B. A history of head injury with dilute urine accompanied by an elevated serum osmolarity suggests central diabetes insipidus. The response of the kidney to exogenous ADH (dDAVP) eliminates nephrogenic diabetes insipidus as the cause of her concentrating defect.

95–E. A negative value for C_{H_2O} means that "free water" (generated in the diluting segments of the thick ascending limb and early distal tubule) is reabsorbed by the collecting ducts. A negative C_{H_2O} is consistent with high circulating levels of ADH. Since ADH levels are high at a time when the serum is very dilute, ADH has been secreted "inappropriately."

96–C. Lithium inhibits the G protein that couples the ADH receptor to adenylate cyclase. The result is inability to concentrate the urine. Because the defect is in the target tissue for ADH (nephrogenic diabetes insipidus), exogenous ADH administered by nasal spray will not correct this defect.

97–D. The description is of a normal person who is deprived of water. The serum osmolarity is slightly above normal because insensible water loss is not being replaced. The rise in serum osmolarity stimulates (via osmoreceptors in the anterior hypothalamus) the release of ADH from the posterior pituitary. ADH then circulates to the kidney and stimulates water reabsorption from the collecting ducts to concentrate the urine.

98–D. Spironolactone inhibits distal tubule Na^+ reabsorption and K^+ secretion by acting as an aldosterone antagonist. Triamterene has similar effects on the distal tubule, but acts directly on the transport processes.

99–A. Acetazolamide, a carbonic anhydrase inhibitor, is used to treat the respiratory alkalosis caused by going to high altitude. It acts on the renal proximal tubule to inhibit the reabsorption of filtered HCO_3^- so that the person excretes an alkaline urine and develops mild metabolic acidosis.

100–B. Thiazide diuretics act on the early distal tubule (cortical diluting segment) to inhibit Na^+ reabsorption. At the same site, they enhance Ca^{2+} reabsorption so that the urinary excretion of Na^+ is increased while the urinary excretion of Ca^{2+} is decreased. K^+ excretion is increased because the flow rate is increased at the site of distal tubular K^+ secretion.

101–B. Gastric parietal cells secrete intrinsic factor, which is required for the intestinal absorption of vitamin B_{12}.

102–A. Saliva has a high $[HCO_3^-]$ because the cells lining the salivary ducts secrete HCO_3^-. Because the ductal cells are relatively impermeable to water and because they reabsorb more solute (Na^+ and Cl^-) than they secrete (K^+ and HCO_3^-), the saliva becomes hypotonic. Vagal stimulation stimulates saliva production, so vagotomy (or atropine) inhibits it and produces dry mouth.

103–C. Secretin stimulates pancreatic secretion, which contains the pancreatic enzymes (i.e., pancreatic lipase, cholesterol ester desmolase, pancreatic phospholipase A_2), which are necessary to convert ingested lipids to an absorbable form.

104–B. When the pH of the stomach contents is very low, the secretion of gastrin by the G cells of the gastric antrum is inhibited. When gastrin secretion is inhibited, then further gastric HCl secretion by the parietal cells is inhibited. Pancreatic secretion is *stimulated* by low pH of the duodenal contents.

105–C. ANF is secreted by the atria in response to ECF volume expansion and subsequently has actions on the kidney that cause increased excretion of Na^+ and H_2O.

106–E. PTH acts on the cells of the renal proximal tubule to activate adenylate cyclase and generate cyclic AMP. Cyclic AMP, through activation of protein kinase, causes inhibition of Na^+– phosphate cotransport in the proximal tubular cells and inhibits phosphate reabsorption.

107–A. ADH not only produces increased water reabsorption in renal collecting ducts (V_2 receptors), but it also causes constriction of vascular smooth muscle (V_1 receptors).

108–A. Angiotensin II increases the production of aldosterone by stimulating aldosterone synthase that catalyzes the conversion of corticosterone to aldosterone.

109–D. 17,20-Lyase catalyzes the conversion of glucocorticoids to the androgenic compounds dehydroepiandrosterone and androstenedione. These androgenic compounds are the precursors of testosterone both in the adrenal cortex and in the testicular Leydig cells.

110–B. Aromatase catalyzes the conversion of testosterone to estradiol in the ovarian granulosa cells. Estradiol is required for female secondary sex characteristics.

111–E. 5α-Reductase catalyzes the conversion of testosterone to dihydrotestosterone. Dihydrotestosterone is the active androgen in the male accessory sex tissues such as prostate.

112–B. The oxidation of I^- to I_2 is inhibited by propylthiouracil and can be used in the treatment of hyperthyroidism.

113–D. The coupling of two molecules of DIT results in the formation of T_4. Coupling of DIT and MIT produces T_3.

114–A. The I^- active transport pump ("trap") in the thyroid follicular cells is inhibited by thiocyanate and perchlorate; these anions can be used to block the incorporation of I^- into thyroid hormones. Radioactive I^- can also be used to destroy the thyroid because it is transported by this pump.

115–E. 5′-Iodinase catalyzes the conversion of T_4 to T_3 (an activation step) in the target tissues for thyroid hormone.

Index

Note: Page numbers followed by f indicate figures; those followed by t indicate tables; those followed by Q indicate questions; and those followed by E indicate explanations.

A

A band, 17, 18f
Absolute refractory period (ARP), 70, 70f
Absorption, 199–205
  alanine, 207Q, 209E
  carbohydrates, 200, 200t, 201f
  cholesterol, 207Q, 209E
  dipeptides, 207Q, 209E
  fatty acid, 207Q, 209E
  fructose, 207Q, 209E
  glucose, 201, 201f, 206Q
  lipids, 200t, 203
  micelle formation for, 207Q, 209E
  NaCl, 203–204
  tripeptides, 207Q, 209E
  vitamin D, 207Q, 209E
  vitamin K, 207Q, 209E
  water, 204
Acetazolamide, uses of, 167t, 262Q, 273E
Acetoacetate, 232
Acetylcholine (ACh), 258Q, 269E
  atrial contractility decrease due to, 74
  degradation of, 14–15
  effects on heart rate, 71
  effects on H^+ secretion, 194f, 195
  in gallbladder contraction, 199
  in mediation of slowing of heart,
    101Q, 105E
  muscarinic receptors for, atropine in
    blockade of, 32, 58Q, 60E
  in regulation of pancreatic secretion,
    198
  synthesis and storage of, 13–15, 14t,
    25Q, 27E
Acetylcholinesterase (AChE), in degrada-
    tion of ACh, 14
Acetylcholinesterase (AChE) inhibitors,
    14, 15, 23Q, 26E
ACh. *See* Acetylcholine
Achalasia, 186
Acid–base balance, 160–166
  buffers, 160–161
  disorders of, 164–166, 164t, 165t,
    171Q, 176E
  in distal potassium secretion, 152f,
    153, 169Q, 174E
  renal, 161–163, 162f–164f
    excretion of H^+ as $NH_4{}^+$, 163, 164f
    excretion of H^+ as titratable acid,
      163, 163f
    reabsorption of filtered $HCO_3{}^-$,
      161–163, 162f
  titration curves, 161, 162f
Acidosis
  effect on distal potassium secretion,
    152f, 153, 169Q, 174E
  metabolic, 164, 164t, 165t, 166, 169Q,
    174E, 235
    causes, 172Q, 176E
  respiratory, 164t, 165t, 166
    COPD and, arterial blood values in,
      172Q, 176E
Acinus
  in formation of pancreatic secretion, 197
  in saliva formation, 191
ACTH. *See* Adrenocorticotropic hormone
Action potential, 11–13, 12f, 23Q, 26E

  cardiac, 67–69, 68f, 69f
    in ventricle, phases of, 68–69, 68f,
      258Q, 268E
  definitions related to, 11
  nerve, 24Q, 27E
    ionic basis of, 11–12, 12f
  overshoot of, 11
  propagation of, 13
  refractory periods of, 11–12, 12f
  repolarization of, 12
  in sinoatrial node, phases of, 69
  undershoot of, 12
  upstroke of, 11, 259Q, 270E
Active tension, 20
Addison's disease, ACTH secretion in,
    247Q, 250E
Adenosine
  in coronary circulation, 91
  exercise effects on, 94
Adenosine triphosphatase (ATPase),
    Na^+-K^+, in cardiac action poten-
    tials, 67
Adenosine triphosphate (ATP), 4, 25Q, 27E
Adenylate cyclase, mechanism of action
    of, 214, 215f
ADH. *See* Antidiuretic hormone
Adrenal cortex, 225–231
  pathophysiology, 230–231
  secretory products of, 227f
Adrenal medulla, 30
  secretion of epinephrine by, cholinergic
    nicotinic receptors in, 58Q, 60E
  secretory products of, 227f
β_2-Adrenergic agonists, for bronchiolar ob-
    struction in asthma, 128Q, 131E
Adrenergic receptors, 30–31, 33t, 57Q,
    58Q, 60E
Adrenocortical hormones
  congenital abnormality due to, 231
  excess of, 230–231
  insufficiency of, 139–140, 139t, 230
    effects of, 247Q, 250E
  secretion of, regulation of, 226–229, 229f
  synthesis of, 225–226, 228f
Adrenocorticotropic hormone (ACTH),
    217, 218f
  decreased, adrenocortical insufficiency
    due to, 230
  described, 260Q, 270E
  increase in, Addison's disease and,
    247Q, 250E
  overproduction of, effects, 247Q, 250E
  POMC in, 246Q, 249E
  in regulation of glucocorticoid secre-
    tions, 227
Adrenogenital syndrome, 231
Adrenoreceptors, 30–31, 33t, 57Q, 58Q,
    60E, 192
Afterload, 20, 20f, 74
  effects on cardiac oxygen
    consumption, 79
Afterpotential, hyperpolarizing, 12
Age
  effect on capacitance of arteries, 66
  effects on pulse pressure, 66
  as factor in glomerular filtration rate, 141
Air flow, 112

  during breathing, 113–114, 113f
Airway obstruction, ventilation/perfusion
    ratio in, 123
Airway resistance, 112–113, 129Q, 132E
  air flow and, 112
  in bronchi, 129Q, 132E
  changes in, factors affecting, 112–113
  Poiseuille's law in, 112
  sites of, 113
Alanine
  absorption of, 207Q, 209E
  reabsorption of, 173Q, 177E
Albright's hereditary osteodystrophy, 237
Albumin, radioactive, in interstitial fluid
    volume measurement, 170Q, 175E
Aldosterone
  actions of, 160t, 230
  blood loss effects on, 94t, 100Q, 105E
  decrease in blood volume effects on,
    100Q, 105E
  in distal potassium secretion,
    152–153, 152f, 169Q, 174E
  H^+ secretion, 150
  in reabsorption of NaCl, 204
  secretion of
    increase in, 160t, 248Q, 251E
    regulation of, 227–229
  synthesis of, 225
  in synthesis of new protein, 247Q, 250E
Aldosterone synthase, 263Q, 273E
Alkaline tide, 193
Alkalosis
  effect on distal potassium secretion,
    152f, 153, 169Q, 174E
  metabolic, 164t, 165t, 166, 171Q, 176E
  respiratory, 164t, 165t, 166
    arterial blood values in, 172Q, 177E
Alpha-motoneurons, 46
  described, 58Q, 61E
Alpha receptors
  alpha-1, 30–31, 33t, 57Q, 60E
  alpha-2, 31, 33t
  in mediation of constriction of arteriolar
    smooth muscle, 101Q, 105E
Altitude, high
  adaptation to, 127, 127t, 128Q, 132E
  ascent to, hyperventilation due to,
    258Q, 269E
  residing at, hypoxemia due to, 129Q,
    132E, 257Q, 268E
Alveolar pressure
  during expiration, 114
  at rest, 113, 113f
Alveolar ventilation, 108
Alveolus(i), surface tension of, 110–111,
    111f
Amacrine cells, 37
Amenorrhea, prolactin excess and, 221
Amiloride, in distal potassium secretion,
    153, 167t, 169Q, 174E
Amine hormone, synthesis of, 213
Amino acids
  concentration of, blood, decrease in,
    insulin and, 234
  free, absorption of, 200t, 202, 202f
  transport of, 206Q, 208E